Advanced Courses in Mathematics
CRM Barcelona

Institut d'Estudis Catalans
Centre de Recerca Matemàtica

Managing Editor:
Manuel Castellet

William G. Dwyer
Hans-Werner Henn

Homotopy Theoretic Methods in Group Cohomology

Springer Basel AG

Authors' addresses:

William G. Dwyer
Department of Mathematics
University of Notre Dame
Notre Dame, IN 46556
USA

e-mail: dwyer.1@nd.edu

Hans-Werner Henn
Département de Mathématiques
UFR de Mathématiques et Informatique
7 rue René Descartes
67084 Strasbourg Cedex
France
e-mail: henn@irma.u-strasbg.fr

2000 Mathematical Subject Classification 20J05; 20J06, 14F35, 55U10

A CIP catalogue record for this book is available from the
Library of Congress, Washington D.C., USA

Deutsche Bibliothek Cataloging-in-Publication Data
Homotopy theoretic methods in group cohomology / William G. Dwyer ; Hans-Werner
Henn. - Boston ; Basel ; Berlin : Birkhäuser, 2001
  (Advanced courses in mathematics - CRM Barcelona)
  ISBN 3-7643-6605-2

ISBN 978-3-7643-6605-6   ISBN 978-3-0348-8356-6 (eBook)
DOI 10.1007/978-3-0348-8356-6

Member of the BertelsmannSpringer Publishing Group
Cover design: Micha Lotrovsky, 4106 Therwil, Switzerland
Printed on acid-free paper produced from chlorine-free pulp. TCF ∞

ISBN 3-7643-6605-2

9 8 7 6 5 4 3 2 1   www.birkhäuser-science.com

Dedicated to
Chuck, Chris, Sarah
Jeanie and Davie

Contents

Cohomology of Groups and Unstable Modules over the Steenrod Algebra

Hans-Werner Henn

Preface

This book consists essentially of notes which were written for an Advanced Course on Classifying Spaces and Cohomology of Groups. The course took place at the Centre de Recerca Mathemàtica (CRM) in Bellaterra from May 27 to June 2, 1998 and was part of an emphasis semester on Algebraic Topology. It consisted of two parallel series of 6 lectures of 90 minutes each and was intended as an introduction to new homotopy theoretic methods in group cohomology.

The first part of the book is concerned with methods of decomposing the classifying space of a finite group into pieces made of classifying spaces of appropriate subgroups. Such decompositions have been used with great success in the last 10–15 years in the homotopy theory of classifying spaces of compact Lie groups and p-compact groups in the sense of Dwyer and Wilkerson. For simplicity the emphasis here is on finite groups and on homological properties of various decompositions known as centralizer resp. normalizer resp. subgroup decomposition. A unified treatment of the various decompositions is given and the relations between them are explored. This is preceeded by a detailed discussion of basic notions such as classifying spaces, simplicial complexes and homotopy colimits.

A second crucial ingredient for the progress in the homotopy theory of classifying spaces of compact Lie groups and p-compact groups came from the theory of unstable modules over the Steenrod algebra, in particular the theory of Lannes' functor T_V. The second part discusses how this theory advances our understanding of the mod-p cohomology ring H^*BG of (a suitable class of) groups G and leads in favorable cases even to complete calculations. After recalling some classical results in cohomology of groups, unstable modules are introduced, Lannes' theory is outlined and his calculation of $T_V H^*BG$ explained. Quillen's theory of F-isomorphisms is shown to be equivalent to Lannes' calculation of $T_V H^*BG$ in degree 0. A major theme of this second part is then to exploit the information given by the full computation of $T_V H^*BG$. Furthermore it is shown how partial knowledge of $T_V H^*BG$ can be used to give various approximations to H^*BG.

There is some overlap between the two parts in so far as centralizer decompositions arise naturally in the approximations to H^*BG discussed in the last two sections of part II. Nevertheless the two parts remain logically independant.

We would like to thank the participants of the course for being such a stimulating audience. Our thanks go also to the staff of the CRM and the topologists at the Universitat Autònoma de Barcelona and in particular to Carles Broto for organizing this course and making it such a pleasant experience.

January 2001   William G. Dwyer, Hans-Werner Henn

Classifying Spaces and Homology Decompositions

W.G. Dwyer

Abstract. Suppose that G is a finite group. We look at the problem of expressing the classifying space BG, up to mod p cohomology, as a homotopy colimit of classifying spaces of smaller groups. A number of interesting tools come into play, such as simplicial sets and spaces, nerves of categories, equivariant homotopy theory, and the transfer.

1. Introduction

In these notes we discuss a particular technique for trying to understand the classifying space BG of a finite group G. The technique is especially useful for studying the cohomology of BG, but it can also serve other purposes. The approach is pick a prime number p and construct BG, up to mod p cohomology, by gluing together classifying spaces of proper subgroups of G. A construction like this is called a *homology decomposition* of BG; in principle it gives an inductive way to obtain information about BG from information about classifying spaces of smaller groups. These homology decompositions are certainly interesting on their own, but another reason to work with them is that it illustrates how to use some everyday topological machinery. We try to make the machinery easier to understand by explaining some things that are usually taken for granted.

The outline of the paper is this. Section 2 introduces classifying spaces and shows how to construct them. Section 3 faces the issue that for our purposes it is much easier to work with combinatorial models for topological spaces than with topological spaces themselves. There is an extended attempt to motivate the particular combinatorial models we will use, called *simplicial sets*; the section ends with a description of the simplicial sets that correspond to classifying spaces. Section 4 gives a systematic account of a large class of gluing constructions called *homotopy colimits*. The easiest way to describe homotopy colimits is with the help of *simplicial spaces* (although from the point of view

Received by the editors June 13, 2001.

of §3 simplicial spaces look a little peculiar!). Simplicial spaces lead to an appealing spectral sequence for the homology of a homotopy colimit (4.16). Section 5 focuses on the *nerve* of a category; this is the homotopy colimit of the constant functor with value a one-point space. It turns out that there is a way using the "Grothendieck Construction" to represent other more complex homotopy colimits as nerves. The advantage of this is that it is easy (by using natural transformations between functors) to construct homotopies between maps of nerves, while constructing combinatorial homotopies between maps of arbitrary simplicial sets is a tedious and error-prone business. Section 6 introduces another specialized homotopy colimit (the *homotopy orbit space*) associated to a group action.

In §7, at last, the main characters come on stage: there is a definition of *homology decomposition* in terms of homotopy colimits, and an explanation (using the results of §5) of how any collection of subgroups of G which satisfies a certain "ampleness" property (7.7) gives rise to three distinct homology decompositions for BG. Section 8 points out that some homology decompositions, called *sharp* decompositions, are better than others, in that a sharp decomposition gives a formula for the homology of BG instead of just a spectral sequence. In this section there are also some examples of homology decompositions, although only a few trivial examples are analyzed in full. The rest of the paper is devoted to working out some nontrivial cases. Section 9 shows that the spectral sequence associated to a homology decomposition can be interpreted as an "isotropy spectral sequence" associated to the action of G on a space X. The following sections (§10, §11) give a homological interpretation of the E^2-page of the isotropy spectral sequence, and develop methods for proving that this E^2-page collapses in such a way that $\mathrm{H}_* \, \mathrm{B}G$ appears along the vertical axis and all other groups are zero. If such a collapse takes place, the corresponding homology decomposition in sharp. Sections 12 and 13 apply this homological machinery to study two specific collections of subgroups of G, the collection of all nontrivial p-subgroups, and the collection of all nontrivial elementary abelian p-subgroups. It turns out that both collections are ample (as long as p divides the order of G), and that many of the six homology decompositions associated to these collections are sharp. The appendix contains a result on G-spaces which is referred to in §10.

2. Classifying spaces

Let G be a discrete group. Later on, we will assume that G is finite.

2.1. Definition. A *classifying space* for G is a pointed connected CW-complex B such that $\pi_1 B$ is isomorphic to G and $\pi_i B$ is trivial for $i > 1$.

2.2. *Remark.* We will usually assume that a classifying space B for G comes with a chosen isomorphism $\iota_B : \pi_1 B \approx G$.

2.3. *Example.* The circle S^1 is a classifying space for the infinite cyclic group $\mathbb{Z}$; real projective space $\mathbb{R}P^\infty$ is a classifying space for $\mathbb{Z}/2$. Let $\mathbb{Z}/p^k$ act on the unit sphere S^{2n-1} in $\mathbb{C}^n$ in such a way that the generator acts by multiplying each coordinate by $e^{2\pi i/k}$. The infinite lens space $X = \cup_n S^{2n-1}/(\mathbb{Z}/p^k)$ is a classifying space for $\mathbb{Z}/p^k$. If X is a classifying space for G and Y is a classifying space for K, then $X \times Y$ is a classifying space for $G \times K$.

2.4. Theorem. *Any discrete group G has a classifying space.*

Sketch of proof. Start with a presentation of G by generators $\{x_\alpha\}$ and relations $\{r_\beta\}$. Let B_0 be a one-point space, and B_1 a space obtained from B_0 by attaching a 1-cell e_α for each x_α, so that B_1 is a bouquet of circles and $\pi_1 B_1$ is isomorphic to the free group $F(x_\alpha)$ on the symbols $\{x_\alpha\}$. Construct B_2 from B_1 by attaching a 2-cell e_β for each relation r_β; the attaching map for e_β should be the word in $F(x_\alpha)$ given by r_β. By the van Kampen theorem, $\pi_1 B_2$ is isomorphic in a natural way to G. Now by induction build B_n from B_{n-1} $(n > 2)$ by attaching n-cells to B_{n-1} via maps $\{f_\gamma : S^{n-1} \to B_{n-1}\}$ which run through a collection of generators for $\pi_{n-1} B_{n-1}$. It is not hard to see that $\pi_1 B_n \approx G$ and $\pi_i B_n \approx 0$ for $2 \le i \le n-1$. Let $B = \cup_n B_n$. Since the sphere S^k is compact, any map $f : S^k \to B$ has image contained in B_n for some n, and so f is null homotopic for $k \ge 2$. It follows from this that B is a classifying space for G. $\square$

> With minor adjustments, the above method can be used to make more general constructions. Let X be a pointed space, W a pointed finite CW-complex, and $\Sigma^k W = S^k \wedge W$ the k-fold suspension of W. Let $X_1 = X$ and by induction build X_n $(n \ge 2)$ from X_{n-1} by attaching a copy of a cone on $\Sigma^k W$ for each map $\Sigma^k W \to X$ $(k \ge 0)$. Let $X_\infty = \cup X_k$. Then any map $f : \Sigma^k W \to X_\infty$ has image contained in X_n for some n, and so f is null homotopic; in fact X_∞ is in a certain sense universal with respect to this mapping property [4]. The space X_∞ is denoted $P_W(X)$. One can remove the requirement that W be a finite complex by iterating the process transfinitely. One further generalization (replacing W by a map $f : U \to V$) yields a functor L_f; essentially all known homotopy theoretic localization functors are of the form L_f for suitable f [5, 10].

For the rest of this section we will let BG denote some chosen classifying space for G.

2.5. Proposition. *If X is any classifying space for G, then up to homotopy there is a unique basepoint-preserving map $f : BG \to X$ such that $\iota_{BG} = \iota_X \cdot f_\#$ (cf. 2.2). The map f is a homotopy equivalence.*

Sketch of proof. Let f send the basepoint of BG to the basepoint of X and each 1-cell e_α of BG to a loop in X representing $(\iota_X)^{-1}\iota_{BG}(\langle e_\alpha\rangle)$. Now extend over the 2-skeleton of BG by using the fact that any relation among the homotopy classes $\{\langle e_\alpha \rangle\}$ in $\pi_1 BG$ also holds among their images in $\pi_1 X$, and extend over higher skeleta of BG by using the fact that the higher homotopy groups of X are trivial. The same idea gives a basepoint-preserving homotopy between f and any other map f' of the same type. The map f is obviously a weak equivalence (i.e. induces isomorphisms $\pi_i BG \approx \pi_i X$); since the domain and range of f are CW-complexes, f is a homotopy equivalence.   □

A similar argument shows that if Y is any pointed connected CW-complex, then the space of pointed maps $Y \to BG$ is weakly equivalent to the discrete set of homomorphisms $\pi_1 Y \to G$. What is the homotopy type of the space of *all* maps $Y \to BG$?

Suppose that X is a pointed, connected CW-complex, and that $\tilde{X}$ is its universal cover. The group $G = \pi_1(X)$ acts freely on $\tilde{X}$ (say on the left) by covering transformations. Suppose that M is a right G-module and that N is a left G-module. For any space Y, let $S_*(Y)$ denote the singular chain complex of Y. Observe that $S_*(\tilde{X})$ is a chain complex of *free* modules over $\mathbb{Z}[G]$; this follows from the fact that if σ is a singular simplex of X, the singular simplices of $\tilde{X}$ which lift σ are permuted freely and transitively by G.

2.6. Definition. The *homology of X with coefficients in M*, denoted $H_*(X; M)$, is the homology of the chain complex $M \otimes_{\mathbb{Z}[G]} S_*(\tilde{X})$. The *cohomology of X with coefficients in N*, denoted $H^*(X; N)$ is the cohomology of the cochain complex $\mathrm{Hom}_{\mathbb{Z}[G]}(S_*(\tilde{X}), N)$.

These are sometimes called *twisted* or *local coefficient* (co-)homology groups of X. If $\mathbb{Z}$ is a trivial G-module, there is a natural isomorphism

$$\mathbb{Z} \otimes_{\mathbb{Z}[G]} S_*(\tilde{X}) \approx S_*(X) .$$

If M is an arbitrary trivial G-module, the resulting isomorphisms

$$M \otimes_{\mathbb{Z}[G]} S_*(\tilde{X}) \approx M \otimes_{\mathbb{Z}} \mathbb{Z} \otimes_{\mathbb{Z}[G]} S_*(\tilde{X}) \approx M \otimes_{\mathbb{Z}} S_*(X)$$

show that $H_*(X; M)$ as defined above agrees with the usual singular homology of X with coefficients in M. A similar result holds for cohomology.

The antiautomorphism of G sending g to g^{-1} gives a way of passing from right G-modules to left G-modules, and vice versa. For this reason the distinction between right G-modules and left G-modules is not too important, and from now on we will not always pay attention to it.

If $X = BG$, then $\tilde{X}$ is contractible, so $S_*(\tilde{X})$ is a free resolution of the trivial G-module $\mathbb{Z}$ over $\mathbb{Z}[G]$ [23, V.11]. In this case elementary homological algebra gives formulas for the homology and cohomology groups of X.

2.7. Proposition. *Suppose that M is a right G-module and that N is a left G-module. Then there are natural isomorphisms*

$$H_*(BG; M) \approx \mathrm{Tor}_*^{\mathbb{Z}[G]}(M, \mathbb{Z})$$
$$H^*(BG; N) \approx \mathrm{Ext}_{\mathbb{Z}[G]}^*(\mathbb{Z}, N)$$

> For a general connected CW-complex X with fundamental group G there is a first-quadrant spectral sequence converging to $H_*(X; M)$ whose E_2-page depends in an algebraic way on the group G and the G-modules $H_*(\tilde{X}; \mathbb{Z})$ and M. Can you construct this spectral sequence and identify its E_2-page?

2.8. *Remark.* The twisted homology $H_*(BG; M)$ (resp. the twisted cohomology $H^*(BG; N)$) is sometimes denoted $H_*(G; M)$ (resp. $H^*(G; N)$) and called the *group homology* of G with coefficients in M (resp. the *group cohomology* of G with coefficients in N). There is a small risk of confusing $H_*(G; M)$ and $H^*(G; M)$ with the homology or cohomology of the discrete space G, but usually this is not a problem. The groups $H_*(G; M)$ have an algebraic interpretation; they are (2.7) the left derived functors of the construction which assigns to any G-module M the abelian group $M' = M \otimes_{\mathbb{Z}[G]} \mathbb{Z}$, i.e., the largest quotient M' of M which is a trivial G-module. Similarly, $H^*(G; N)$ is the collection of right derived functors of the construction $N \mapsto N^G = \mathrm{Hom}_{\mathbb{Z}[G]}(\mathbb{Z}, N)$.

> Let G be a finitely generated abelian group. Starting from 2.3, compute $H^*(BG; \mathbb{F}_p)$ and $H^*(BG; \mathbb{Q})$ as rings. A more ambitious project is to compute the homology and cohomology of $B\mathbb{Z}_{p^\infty}$, where $\mathbb{Z}_{p^\infty} = \cup_n \mathbb{Z}/p^n$. The rational homology of this space is trivial, but the integral cohomology is torsion-free. What is the rational cohomology?

2.9. *Remark.* Fix a prime number p, and let $\mathbb{F}_p$ denote the field with p elements. Let G be a finite group. In these notes we will construct various maps $f : A \to BG$ which approximate BG up to (co-)homology, in the sense that f induces isomorphisms on $H_*(-; \mathbb{F}_p)$ or $H^*(-; \mathbb{F}_p)$.

> One could also look for maps $f : A \to BG$ which induce isomorphisms on $H_*(-; M)$ or on $H^*(-; N)$ for suitable $\mathbb{F}_p[G]$-modules M and N. Let G be a finite group, X a CW-complex with a map $h : \pi_1(X) \to G$, and X' the regular covering space of X corresponding to $\ker(h)$. Show that $H_*(X; \mathbb{F}_p[G])$ is isomorphic to $H_*(X'; \mathbb{F}_p)^{\oplus n}$, where n is the index of $\mathrm{im}(h)$ in G. Conclude that $f : A \to BG$ induces an isomorphism $H_*(A; M) \to H_*(BG; M)$ for all $\mathbb{F}_p[G]$-modules M if and only if $\pi_1(f)$ is surjective and the homotopy fibre of f has the $\mathbb{F}_p$-homology of a point. Is it necessary to assume that G is finite? What happens if f induces an isomorphism $H^*(BG; N) \to H^*(A; N)$ for all $\mathbb{F}_p[G]$-modules N? What happens if BG is replaced by another space Y?

2.10. *Notation.* From now on, all homology and cohomology will have coefficients in $\mathbb{F}_p$, unless some other coefficients are specified. If there is a fundamental group G involved, we will always assume that $\mathbb{F}_p$ has a trivial G-action.

3. Simplicial complexes and simplicial sets

In order to make progress on building homology approximations to BG (2.9), we need to develop a method of describing spaces combinatorially. CW-complexes have nice topological properties, but without additional machinery it is not easy to give explicit cellular recipes for complicated spaces.

Abstract simplicial complexes. An *abstract simplicial complex* K is a pair (V_K, S_K), where V_K is a set and S_K is a collection of nonempty finite subsets of V_K. The elements of V_K are the *vertices* of K and the elements of S_K are the *simplices* of K. The collection S_K is required to be closed under passage to subsets: if $\sigma \in S_K$ and $\sigma' \subset \sigma$, then $\sigma' \in S_K$. To avoid carrying around extra baggage, we also insist that if $v \in V_K$ the singleton subset $\{v\}$ should belong to S_K. The subsets of a simplex σ are called its *faces*.

Associated to an abstract simplicial complex K is a space $|K|$, called the *geometric realization* of K, obtained as the space of formal linear combinations

$$(3.1) \qquad \qquad \sum_{v \in V_K} t_v v \,,$$

where $0 \le t_v \le 1$, $\mathrm{supp}\{t_v\} \in S_K$, and $\sum t_v = 1$. (Here $\mathrm{supp}\{t_v\}$, called the support of $\{t_v\}$, is the set of all vertices v such that $t_v \ne 0$). The topology on $|K|$ is given as follows [26, §1.1–3]: for each simplex $\sigma \in S_K$ the subspace $\sum_{v \in \sigma} t_v v$ is topologized as a subspace of $\mathbb{R}^{\mathrm{card}(\sigma)}$, and then $|K|$ is given the finest topology with respect to which the inclusions of these subspaces are continuous. A simplex $\sigma \in S_K$ with $\mathrm{card}(\sigma) = n + 1$ is called an n-simplex of K, because the subspace of $|K|$ corresponding to σ is a topological n-simplex.

Abstract simplicial complexes form the objects of a category **ASC**, in which a map $K \to L$ is a map of sets $f : V_K \to V_L$ such that for each simplex σ of K, $f(\sigma)$ is a simplex of L. Such a map induces a continuous map $f : |K| \to |L|$ according to the formula

$$f\left(\sum t_v v\right) = \sum t_v f(v) \,.$$

We do not require f to be a monomorphism, so it might be necessary to rearrange terms on the right hand side of the above equality and add up the coefficients of any given vertex of L in order to obtain a sum in standard form (3.1).

3.2. *Examples.* The *abstract n-simplex* $D[n]$ is the abstract simplicial complex whose vertex set is $\mathbf{n} = \{0, 1, \dots, n\}$ and whose collection of simplices is the set of all subsets of $\mathbf{n}$. The geometrical realization $|D[n]|$ is a standard topological n-simplex. The automorphism group of $D[n]$ is the symmetric group Σ_{n+1}, which acts by permuting vertices. There is a unique map $D[n] \to D[0]$, which on taking geometrical realizations gives the map from a topological n-simplex to a one-point space.

Describing spaces in terms of abstract simplicial complexes has a direct geometric appeal [32, §3] [26, §1], but for a couple of reasons it is difficult to use the category of abstract simplicial complexes as a foundation for homotopy theory. One problem is that the categorical product in **ASC** is not well-behaved, in the sense that it does not commute with geometric realization.

> If X and Y are objects of some category $\mathbf{C}$, an object Z of $\mathbf{C}$ is called a *(categorical) product* of X and Y if Z is provided with morphisms (called projection maps) $\mathrm{pr}_1 : Z \to X$ and $\mathrm{pr}_2 : Z \to Y$ such that for any object W of $\mathbf{C}$, composition with these maps induces a bijection
>
> $$\mathrm{Hom}_{\mathbf{C}}(W, Z) \to \mathrm{Hom}_{\mathbf{C}}(W, X) \times \mathrm{Hom}_{\mathbf{C}}(W, Y) \ .$$
>
> It is easy to check that if Z and Z' are two products for X and Y, then there is a unique isomorphism $Z \to Z'$ which is compatible with the projection maps, and so it is common to speak of "the" product of X and Y. See [22, III.4] for an extended discussion of products and, more generally, limits.

Suppose that K and L are abstract simplicial complexes. Let $V_{K \otimes L}$ denote the cartesian product $V_K \times V_L$, let $\mathrm{pr}_1 : V_{K \otimes L} \to V_K$ and $\mathrm{pr}_2 : V_{K \otimes L} \to V_L$ be the (set theoretic) projection maps, and let $S_{K \otimes L}$ denote the collection of all subsets σ of $V_{K \otimes L}$ such that $\mathrm{pr}_1(\sigma) \in V_K$ and $\mathrm{pr}_2(\sigma) \in V_L$.

3.3. Lemma. *In the above situation, the abstract simplicial complex $K \otimes L = (V_{K \otimes L}, S_{K \otimes L})$ is the categorical product of K and L in* **ASC**. *In general, the natural map $|K \otimes L| \to |K| \times |L|$ is not a homeomorphism.*

Proof. The first statement in the lemma is obvious. For the second, it's only necessary to notice that $D[1] \otimes D[1]$ contains the 3-simplex $\{(0,0), (0,1), (1,0), (1,1)\}$. □

The example $D[1] \otimes D[1]$ suggests that the categorical product $K \otimes L$ of two abstract simplicial complexes is too big because it is forced to be a target of too many maps. One way to deal with this problem is to reduce the number of maps in the category by introducing some extra structure on the objects. It turns out that a good way to do this is to introduce a total ordering on the vertex set of each simplex.

Ordered simplicial complexes. An *ordered simplicial complex* K is an abstract simplicial complex (V_K, S_K) with the property that V_K is furnished with a partial ordering [32, 1.1] which restricts to a total ordering on each simplex of K. The geometric realization $|K|$ of an ordered simplicial complex is defined to be the geometric realization of the underlying simplicial complex. Ordered simplicial complexes form the objects of a category **OSC**, in which the morphisms are the maps $K \to L$ of simplicial complexes which respect the partial orderings on the vertex sets.

3.4. *Remark.* A set with a partial ordering is called a *partially ordered set*, or *poset* for short.

3.5. *Examples.* Any abstract simplicial complex can be converted to an ordered simplicial complex by choosing a total ordering on its set of vertices. For instance, the *ordered n-simplex* $\Delta[n]$ is the ordered simplicial complex whose vertex set is the set $\mathbf{n}$ (3.2) with the usual numerical ordering and whose collection of simplices is the set of all nonempty subsets of $\mathbf{n}$. The geometrical realization $|\Delta[n]|$ is a standard topological n-simplex. The automorphism group of $\Delta[n]$ is trivial, since there are no nonidentity order-preserving bijections $\mathbf{n} \to \mathbf{n}$. As in 3.2, there is a unique map $\Delta[n] \to \Delta[0]$ which corresponds under geometric realization to the map from a topological n-simplex to a one-point space.

3.6. *Example.* Suppose that S is a poset. Associated to S is a "largest possible" ordered simplicial complex $\mathrm{Cx}(S)$ with vertex poset S; this is the complex whose simplices consist of *all* of the totally ordered subsets of S. Given any simplicial complex K, ordered or not, let $S = S_K$ be the set of simplices of K, ordered under inclusion. The ordered simplicial complex $\mathrm{Cx}(S_K)$ is called the *barycentric subdivision* [32, p. 123–124] of K, and denoted $\mathrm{sd}\,K$. The geometric realization of $\mathrm{sd}\,K$ is homeomorphic to the geometric realization of K.

Suppose that K and L are ordered simplicial complexes. Let $V_{K \times L}$ denote the cartesian product $V_K \times V_L$ with the product partial ordering (i.e., $(v, w) \leq (v', w')$ if $v \leq v'$ and $w \leq w'$). Let $\mathrm{pr}_1 : V_{K \times L} \to V_K$ and $\mathrm{pr}_2 : V_{K \times L} \to V_L$ be the projection maps, and let $S_{K \times L}$ denote the collection of all totally ordered subsets σ of $V_{K \times L}$ such that $\mathrm{pr}_1(\sigma) \in S_K$ and $\mathrm{pr}_2(\sigma) \in S_L$.

The next proposition shows that replacing the category **ASC** by **OSC** solves the problem of products.

3.7. Proposition. *In the above situation, the ordered simplicial complex $K \times L = (V_{K \times L}, S_{K \times L})$ is the product of K and L in the category of ordered simplicial complexes. The natural map $|K \times L| \to |K| \times |L|$ is a homeomorphism.*

Proof. The first statement in the lemma is obvious. The indicated natural map takes a point $\sum c_k(v_k, w_k)$ of $|K \times L|$ to the point

$$\left(\sum c_k v_k, \sum c_k w_k \right)$$

of $|K| \times |L|$. The inverse homeomorphism $|K| \times |L| \to |K \times L|$ takes

$$\left(\sum s_i v_i, \sum t_j w_j \right)$$

to the point $\sum c_k p_k$ determined as follows. Relabel so that the vertices of K and L which appear with a nonzero coefficient are

$$v_0 < v_1 < \cdots < v_m \quad \text{and} \quad w_0 < w_1 < \cdots < w_n \ .$$

Let $S_i = \sum_{i' \leq i} s_{i'}$ and $T_j = \sum_{j' \leq j} t_{j'}$. Note that $S_m = T_n = 1$. Let

$$C_0 \leq C_1 \leq \cdots \leq C_{m+n}$$

be the sequence that results from arranging the $(n + m + 2)$ numbers T_i and S_j in order and deleting one of the repeated 1's at the high end. Then $p_k = (v_{a(k)}, w_{b(k)})$, where $a(k)$ is the cardinality of the set $\{i \mid S_i \leq C_k\}$ and $b(k)$ is the cardinality of $\{j \mid T_j \leq C_k\}$. The coefficient c_k is $C_k - C_{k-1}$. See [15, p. 68].   □

3.8. *Remark.* The best way to understand the above proposition is to look at the triangulations it gives for products of low-dimensional simplices. Start with $|\Delta[1]| \times |\Delta[1]|$ and go on to $|\Delta[1]| \times |\Delta[2]|$. The process is less complicated than it seems to be at first sight. The top-dimensional simplices of $\Delta[p] \times \Delta[q]$ are indexed by (p, q)-*shuffles* [24, p. 17].

3.9. *Remark.* The careful reader will worry about whether the "inverse home-omorphism" described in the proof of 3.7 is continuous. In fact it is *not* contin-uous unless K and L satisfy some finiteness conditions or (which is usually the case in homotopy theory) the product $|K| \times |L|$ is tacitly given the compactly generated topology [33].

Now that we have handled products, we have to deal with a second defi-ciency of the category **ASC**, a deficiency which is inherited by **OSC**; this is the poor behavior of pushouts (or more generally, of colimits).

Given a diagram $Z \xleftarrow{g} X \xrightarrow{f} Y$ in some category **C**, a *pushout* for the diagram is an object P of **C** together with a commutative square (on the left)

$$
\begin{array}{ccc}
X & \xrightarrow{f} & Y \\
{\scriptstyle g}\downarrow & & \downarrow{\scriptstyle u} \\
Z & \xrightarrow{v} & P
\end{array}
\qquad\qquad
\begin{array}{ccc}
X & \xrightarrow{f} & Y \\
{\scriptstyle g}\downarrow & & \downarrow{\scriptstyle u'} \\
Z & \xrightarrow{v'} & P'
\end{array}
$$

such that for any commutative square on the right there is a unique morphism $t : P \to P'$ with $tu = u'$ and $tv = v'$. See [22, p. 66]; any two pushouts for the same diagram are isomorphic in a unique way respecting the pushout structure maps. The intuitive content of this notion is that the pushout P is obtained by gluing Y and Z together along X. See [22, III.1] for a discussion of the more general notion of colimit.

If $L \leftarrow K \to L'$ is a diagram of ordered simplicial complexes, one would hope to be able to form its pushout W in such a way that $|W|$ would be homeomorphic to the pushout of $|L| \leftarrow |K| \to |L'|$. This is impossible. Let $\partial\Delta[n]$ be the subcomplex of $\Delta[n]$ containing all simplices except $\{0, 1, \ldots n\}$. The pushout of the diagram

$$\Delta[0] \leftarrow \partial\Delta[n] \to \Delta[n] .$$

in **OSC** is isomorphic to $\Delta[0]$: just calculate with vertices, and observe that any ordered simplicial complex with one vertex is isomorphic to $\Delta[0]$. The pushout of the corresponding geometric diagram

$$
\begin{array}{ccccc}
|\Delta[0]| & \longleftarrow & |\partial\Delta[n]| & \longrightarrow & |\Delta[n]| \\
\approx\downarrow & & \approx\downarrow & & \approx\downarrow \\
* & \longleftarrow & S^{n-1} & \longrightarrow & D^n
\end{array}
$$

(3.10)

is the n-sphere. The problem here is that a simplex in an ordered simplicial complex is by definition determined by its vertices, so that identifying vertices with one another automatically collapses any simplices which they form. In the topological category, however, it is possible to identify all of the vertices of a simplex to a point, or even to collapse the whole boundary of a simplex to a point, without collapsing the simplex itself.

The solution to this problem is to stop thinking of a simplex as determined by its vertices, and instead to think of it as an indivisible object which has vertices (and faces of other dimensions) associated to it in some explicit way. Of course these vertices (and other faces) are again indivisible objects of the same general type. This leads to the notion of a *simplicial set*.

Simplicial sets. Let K be an ordered simplicial complex. We think of a map $\sigma : \Delta[n] \to K$ as a kind of generalized simplex of K. Note that the maps $\Delta[n] \to K$ which give monomorphisms $\mathbf{n} \to V_K$ correspond bijectively to ordinary simplices of K, but it is a bad idea to restrict attention to these special kinds of maps, because they don't behave functorially.

> What does "they don't behave functorially" mean?

A generalized simplex σ of K corresponds to a face of σ' if $\sigma = \sigma' \cdot i$ for some monomorphism $i : \Delta[n] \to \Delta[m]$. Similarly, σ is obtained from σ' by repeating vertices if $\sigma = \sigma \cdot j$ for some epimorphism $j : \Delta[n] \to \Delta[m]$. This suggests considering the following structure.

3.11. Definition. Let Δ denote the full subcategory of **OSC** with objects $\Delta[n]$ ($n \geq 0$). A *simplicial set* X is a functor

$$X : \Delta^{\mathrm{op}} \to \mathbf{Sets} .$$

> The category Δ^{op} is the *opposite category* of Δ [22, p.33]. A functor with domain Δ^{op} is the same thing as a contravariant functor with domain Δ.

Simplicial sets form a category **Sp** in which a morphism is a natural transformation between functors. If X is a simplicial set, we write X_n for the set $X(\Delta[n])$ and call this the set of *n-simplices* of X. The above considerations provide a "singular" functor $\mathrm{Sing} : \mathbf{OSC} \to \mathbf{Sp}$ given by

$$\mathrm{Sing}(K)_n = \mathrm{Hom}_{\mathbf{OSC}}(\Delta[n], K) .$$

It is easy to see that the category Δ can be identified with the category whose objects are the ordered sets $\mathbf{n}$ ($n \geq 0$) and whose maps are the set maps

$f : \mathbf{n} \to \mathbf{m}$ which are weakly order preserving (in the sense that $f(i) \le f(j)$ if $i \le j$); this amounts to noticing that a map $\Delta[n] \to \Delta[m]$ is completely determined by its effect on vertices. One can then check that the morphisms in Δ are generated by injective maps $d^i : \mathbf{n-1} \to \mathbf{n}$ and surjective maps $s^i : \mathbf{n+1} \to \mathbf{n}$ $(0 \le i \le n)$ given by the following formulas:

$$(3.12) \qquad d^i(j) = \begin{cases} j & \text{if } j < i \\ j+1 & \text{if } j \ge i \end{cases} \qquad\qquad s^i(j) = \begin{cases} j & \text{if } j \le i \\ j-1 & \text{if } j > i \end{cases}$$

These morphisms satisfy a few obvious relations

$$(3.13) \qquad \begin{aligned} d^j d^i &= d^i d^{j-1} \text{ if } i < j \\ s^j s^i &= s^i s^{j+1} \text{ if } i \le j \\ s^j d^i &= d^i s^{j-1} \text{ if } i < j \\ s^j d^j &= \text{identity} = s^j d^{j+1} \\ s^j d^i &= d^{i-1} s^j \text{ if } i > j+1 \end{aligned}$$

and it is possible to verify that all relations between composites of the d^i's and s^j's are consequences of these specific ones [24, p. 4]. This leads to another way to look at a simplicial set X. Recall that X is a contravariant functor on Δ; let $d_j = X(d^j)$ and $s_j = X(s^j)$. Then X is effectively a collection of sets X_n $(n \ge 0)$ together with maps $d_i : X_n \to X_{n-1}$ and $s_i : X_n \to X_{n+1}$ $(0 \le i \le n)$ satisfying a list of identities [24, p. 1] which are the opposites of the ones in (3.13). For completeness, we will write these identities out:

$$(3.14) \qquad \begin{aligned} d_i d_j &= d_{j-1} d_i \text{ if } i < i \\ s_i s_j &= s_{j+1} s_i \text{ if } i \le j \\ d_i s_j &= s_{j-1} d_i \text{ if } i < j \\ d_j s_j &= \text{identity} = d_{j+1} s_j \\ d_i s_j &= s_j d_{i-1} \text{ if } i > j+1 \ . \end{aligned}$$

The d_i's are called *face operators* and the s_i's *degeneracy operators*. If x is an n-simplex of X, then $d_i x$ is an $(n-1)$-simplex of X which is the i'th face of x.

3.15. Remark. More generally, if $\mathbf{C}$ is some category, a *simplicial object* in $\mathbf{C}$ is by definition just a functor $\Delta^{\mathrm{op}} \to \mathbf{C}$ (see 3.11); these simplicial objects form a category, in which the morphisms are the natural transformations of functors. A simplicial set is a simplicial object in **Sets**. More explicitly, a simplicial object X in $\mathbf{C}$ is a collection $\{X_n\}_{n \ge 0}$ of objects of $\mathbf{C}$ together with morphisms $d_i : X_n \to X_{n-1}$ and $s_i : X_n \to X_{n+1}$ $(0 \le i \le n)$ satisfying the identities 3.14.

If **C** is the category of abelian groups (or more generally the category of modules over a ring R), then the category of simplicial objects over **C** is equivalent (via a normalization functor, see 3.21 or [24, 22.4]) to the category of nonnegatively graded chain complexes over **C**. The study of these chain complexes is homological algebra. The idea of "chain complex" does not extend very well to more general categories **C**, but the idea of "simplicial object" does, and replacing chain complexes by simplicial objects gives a way to generalize homological algebra to essentially arbitrary algebraic settings. Following Quillen [27], this generalization is called *homotopical algebra*. Since simplicial sets are closely connected to topological spaces, ordinary homotopy theory can be thought of as the homotopical algebra of the category of sets! Another instance comes up in [29]; here Quillen studies the homotopical algebra of the category of commutative rings.

The geometric realization construction for ordered simplicial complexes extends to a similar construction for simplicial sets, although the description of this extended construction is a little more complicated. Let $\Delta_n = |\Delta[n]|$ be the topological n-simplex, and for each morphism $d^i : \Delta[n] \to \Delta[n+1]$ or $s^i : \Delta[n] \to \Delta[n-1]$ (3.12) denote by the same symbols the induced continuous maps $\Delta_n \to \Delta_{n+1}$ and $\Delta_n \to \Delta_{n-1}$. (The map $d^i : \Delta_n \to \Delta_{n+1}$ is the i'th face inclusion, and $s^i : \Delta_n \to \Delta_{n-1}$ is a linear collapse obtained by pinching together the i'th and $(i+1)$'st vertices.) If X is a simplicial set, the geometric realization $|X|$ is the space obtained from the disjoint union

$$\bigcup_n X_n \times \Delta_n$$

(here X_n is treated as a space with the discrete topology) by making the identifications

(3.16)
$$(d_i x, p) \sim (x, d^i p) \text{ for } (x, p) \in X_n \times \Delta_{n-1}$$
$$(s_i x, p) \sim (x, s^i p) \text{ for } (x, p) \in X_{n-1} \times \Delta_n$$

See [24, III]. A simplex $x \in X_n$ is said to be *degenerate* if $x = s_i x'$ for some i and some $x' \in X_{n-1}$. The space $|X|$ is a CW-complex in which the n-cells correspond to the nondegenerate n-simplices of X [24, 14.1]. If $X = \mathrm{Sing}(K)$ for an ordered simplicial complex K, then the nondegenerate simplices of X are essentially the ordinary simplices of K, and $|X|$ is homeomorphic to $|K|$.

From the categorical point of view, the geometric realization of a simplicial set X is a curious construction; it is built by gluing, but it certainly is not given as a colimit over Δ. The best way to understand it is as a *coend* [22, p. 223], or equivalently as a kind of tensor product over Δ of the covariant functor $\mathbf{n} \mapsto \Delta_n$ with the contravariant functor $\mathbf{n} \mapsto X_n$.

All small limits and colimits exist in the category **Sp** and can be constructed dimensionwise; see [22, p. 111] for limits, and the dual (i.e. opposite-category)

version for colimits. For example, the categorical product $X \times Y$ of two simplicial sets is given by

$$(X \times Y)_n = X_n \times Y_n$$

with simplicial operators $d_i = d_i^X \times d_i^Y$ and $s_i = s_i^X \times s_i^Y$. If $Z \xleftarrow{g} X \xrightarrow{g} Y$ is a diagram of simplicial sets then the pushout P of the diagram exists, and P_n can be calculated as the pushout of the diagram $Z_n \xleftarrow{g_n} X_n \xrightarrow{f_n} Y_n$ in the category of sets.

The category of simplicial sets solves the two problems we encountered above. First of all, for any simplicial sets X and Y the natural map

$$|X \times Y| \to |X| \times |Y|$$

is a homeomorphism [24, 14.3], at least if $|X| \times |Y|$ is given a suitable topology (3.9). This is an extension of the result above for ordered simplicial complexes, since if K and L are objects of **OSC**, there is an isomorphism $\text{Sing}(K) \times \text{Sing}(L) \approx \text{Sing}(K \times L)$. Secondly, the realization functor **Sp** $\to$ **Top** commutes with pushouts, and, even better, with all colimits [24, 16.1] [22, p. 115]. Note how (3.10) works in **Sp**. The pushout of the diagram

$$\text{Sing}(\Delta[0]) \leftarrow \text{Sing}(\partial\Delta[n]) \to \text{Sing}(\Delta[n])$$

in **Sp** is a simplicial set X which has only two nondegenerate simplices, one of dimension 0 and one of dimension n; X is an object of **Sp** which is genuinely new: it does not come from **OSC**. The geometrical realization $|X|$ is obtained from the topological n-simplex Δ_n by collapsing its boundary to a point, and is homeomorphic to the n-sphere.

3.17. *Remark.* From now on we will identify the ordered simplicial complex $\Delta[n]$ with the corresponding simplicial set $\text{Sing}(\Delta[n])$ and call it the *standard n-simplex*. The set $\Delta[n]_k$ of its k-simplices is the set of nondecreasing sequences $\langle i_0, \ldots, i_k \rangle$ of elements of **n**.

3.18. Homotopy and homology of simplicial sets. For the rest of this paper we will refer to simplicial sets as *spaces*, and sets with topologies as *topological spaces*. This is partly for convenience, and partly to emphasize the fact that in almost everything we do here, point-set topology plays no significant role; what is important are the combinatorial patterns in which simplices are matched with one another. The category **Sp** of simplicial sets is a convenient one for making homotopy theoretic manipulations (see for instance [6]).

3.19. *Remark.* We will think of the category of sets as embedded in **Sp** by the functor which assigns to a set S the constant simplicial set (i.e. constant functor $\Delta^{op} \to$ **Sets**) with value S. These simplicial sets are called *discrete spaces*. Equivalently, this functor assigns to S a coproduct of copies of $\Delta[0]$, one copy for each element of S. The geometric realization of this simplicial set is homeomorphic to the set S itself (endowed with the discrete topology).

> The homotopical relationship between **Sp** and **Top** is very close; besides a realization functor **Sp** → **Top**, there is a singular functor Sing : **Top** → **Sp** given by
>
> $$\mathrm{Sing}(X)_n = \mathrm{Hom}_{\textbf{Top}}(\Delta_n, X) \ .$$
>
> Almost by definition, the cellular homology of the CW-complex $|\mathrm{Sing}(X)|$ is the singular homology of X, so it is not surprising that the more or less obvious natural map of spaces
>
> $$|\mathrm{Sing}(X)| \to X$$
>
> is an isomorphism on homology [24, 16.2] or even a weak homotopy equivalence. The realization functor is *left adjoint* [22, p .81] to the singular functor, and this gives a simple explanation for the fact that the realization functor commutes with colimits [22, p. 114–115]. See [17] for an extended discussion of simplicial theory.

3.20. Definition. A map f of spaces is said to be an *equivalence* or *weak equivalence* if the geometric realization of f is a weak equivalence of topological spaces.

> The *homotopy groups* $\pi_* X$ of a space X can be defined either in terms of the ordinary homotopy groups of $|X|$, or by a direct combinatorial formula [24, §3] [24, 26.7]. Needless to say the combinatorial formula is pretty complicated, since so far it has not led to a calculation of $\pi_*(\Delta[n]/\partial\Delta[n])$! A map of spaces is a weak equivalence in the above sense if and only if it is bijective on components and for every choice of basepoint induces an isomorphism on homotopy groups. Try to find a combinatorial description of the set of components of a space X. How about a construction of the fundamental group of X (based at a vertex)?

Before considering the homology of a space, we have to make an algebraic definition.

3.21. Definition. Suppose that A is a simplicial abelian group (3.15). The *normalization* of A, denoted $N(A)$, is the chain complex obtained by defining

$$N(A)_n = A_n/(s_0 A_{n-1} + \cdots + s_{n-1} A_{n-1})$$

and letting $\partial : N(A)_n \to N(A)_{n-1}$ be the quotient map [24, 22.2] induced by the alternating sum of face operators

$$(3.22) \qquad\qquad \sum_{i=0}^{n} d_i : A_n \to A_{n-1} \ .$$

3.23. *Remark.* The homology groups of $N(A)$ are denoted $\mathrm{H}_* N(A)$ or sometimes, for brevity, just $\mathrm{H}_*(A)$. It is possible to form another chain complex $C(A)$ by setting $C(A)_n = A_n$ and letting $\partial : C(A)_n \to C(A)_{n-1}$ be given by formula (3.22) (without passing to any quotient). This change does not make much of a difference: the obvious surjection $C(A) \to N(A)$ is a chain map which induces an isomorphism on homology groups [24, 22.2].

3.24. Definition. If X is a space, let $\mathbb{Z}[X]$ denote the simplicial abelian group obtained by applying to X (dimensionwise) the free abelian group functor from the category of sets to the category of abelian groups. If M is an abelian group, let $M[X]$ denote the simplicial abelian group $M \otimes_{\mathbb{Z}} (\mathbb{Z}[X])$. The *homology of X with coefficients in M*, denoted $H_*(X; M)$ is defined to be the homology of the chain complex $N(M[X])$.

> How would you define $H^*(X; M)$? There is a latent ambiguity in the above notation: if A is a simplicial abelian group, then $H_*(A)$ could mean either the homology of the chain complex $N(A)$ (3.23) or the homology of the chain complex $N(\mathbb{Z}[A])$ (3.24). We promise not to use the second interpretation. Read a little further, and try to sort out how this second interpretation would involve the homology of Eilenberg-MacLane spaces.

3.25. *Remark.* If Y is a subspace of X, the relative homology $H_*(X, Y; M)$ is defined to be the homology of the chain complex

$$N(M[X]/M[Y]) \approx N(M[X])/N(M[Y]) .$$

It is not hard to see that $H_*(X, Y; M)$ is exactly the cellular homology, with coefficients in M, of the pair $(|X|, |Y|)$ of CW-complexes.

> If A is a simplicial abelian group, there is a natural isomorphism [24, 22.1]
>
> $$\pi_*(A) \approx H_*(N(A)) .$$
>
> In particular, if X is a space, $|\mathbb{Z}[X]|$ is space whose homotopy groups are the ordinary (singular or cellular) integral homology groups of $|X|$. Compare this to the Dold-Thom theorem [9], which expresses the homology of a topological space Y as the homotopy of the infinite symmetric product $SP^\infty(Y)$. There is quite a similarity here: $\mathbb{Z}[X]$ is the simplicial free abelian group on X, while $SP^\infty(Y)$ is the topological free abelian monoid on Y.

3.26. Classifying spaces. An action of a (discrete) group G on a space X amounts to actions of G on the sets X_n ($n \geq 0$) which commute with all face and degeneracy operators. An action of G on X is said to be *free* if the induced action of G on each X_n is free. A space X is said to be *weakly contractible* if the unique map $X \to * = \Delta[0]$ is a weak equivalence, or, in other words, if $|X|$ is a contractible topological space. Suppose that E is a weakly contractible space with a free action of G. It is not hard to see that the natural map

$$|E| \to |E/G|$$

is a principal covering map with group G, and so $|E/G|$ is a topological classifying space (2.1) for G. In the combinatorial context we have chosen to work in, we will call E/G itself a classifying space for G. Later on (5.9) we will describe a functorial construction for classifying spaces.

4. Simplicial spaces and homotopy colimits

The construction of the homotopy colimit is motivated by the fact that ordinary colimits are not well-behaved with respect to weak equivalences. For instance, consider the following commutative diagram of topological spaces (where D^n is the n-disk and S^{n-1} its boundary sphere).

(4.1)
$$
\begin{array}{ccccc}
D^n & \longleftarrow & S^{n-1} & \longrightarrow & D^n \\
\downarrow & & =\downarrow & & \downarrow \\
* & \longleftarrow & S^{n-1} & \longrightarrow & *
\end{array}
$$

All three vertical arrows are weak equivalences (even homotopy equivalences) but the colimit of the top row is homeomorphic to S^n, the colimit of the bottom row is a one-point space $*$, and the map $S^n \to *$ induced by the diagram is not a weak equivalence. The same sort of thing can happen with spaces (i.e., simplicial sets (3.18)); consider the diagram

(4.2)
$$
\begin{array}{ccccc}
\Delta[n] & \longleftarrow & \partial\Delta[n] & \longrightarrow & \Delta[n] \\
\downarrow & & \downarrow & & \downarrow \\
* & \longleftarrow & \partial\Delta[n] & \longrightarrow & *
\end{array}
$$

where in this case $* = \Delta[0]$. The lesson from elementary homotopy theory is that the colimit of the top row in (4.1) is the "correct" homotopy pushout: in a homotopical context, before taking the pushout of a diagram of topological spaces you should first replace the maps by equivalent cofibrations. There is a parallel principle in **Sp**; before taking the pushout of a diagram of simplicial sets, you should first replace the maps by weakly equivalent injections. What is gained by such replacement is homotopy invariance (cf. 4.14). The general homotopy colimit construction we will describe below makes these procedures systematic, and gives a way to generalize them to colimits more complicated than pushouts. The result is that we will be able to use a single language to discuss a large family of homotopy invariant ways to glue spaces together.

See [13, §10] for a conceptual approach to homotopy pushouts. From the point of view presented there (which can be traced back to [6] and [27]), the homotopy pushout is a derived functor of the pushout [13, 10.7], in a sense closely related to the way Tor is a derived functor of $\otimes$ [13, 9.6]. There is a similar relationship between general colimits and the corresponding homotopy colimits.

The homotopy colimit is defined in terms of a construction which is also useful for other purposes. This is the construction of the *realization of a simplicial space*.

Simplicial spaces and their realizations. Suppose that X is a simplicial space (3.15), i.e., a simplicial object in the category of spaces.

4.3. Definition. The *realization* of X is defined to be the space constructed by taking the disjoint union

$$(4.4) \qquad\qquad \bigcup_n X_n \times \Delta[n]$$

and making the analogues of identifications (3.16).

4.5. *Remark.* There is another very different way to construct the realization. The object X is a sequence $\{X_n\}$ of spaces, with "horizontal" maps $d_i^h :$ $X_n \to X_{n-1}$, $s_i^h : X_n \to X_{n-1}$. Each X_n is itself a sequence of sets $X_{m,n}$ with "vertical" maps $d_i^v : X_{m,n} \to X_{m-1,n}$, $s_i^v : X_{m,n} \to X_{m-1,n}$. Combining the two directions allows X to be considered as a rectangular array $\{X_{m,n}\}$ of sets with both horizontal and vertical simplicial operators. The following proposition may be hard to believe, but it is easy to check.

4.6. Proposition. *If X is a simplicial space, then the realization of X is naturally isomorphic to the space $\mathrm{diag}(X)$ with $\mathrm{diag}(X)_n = X_{n,n}$. The face operators d_i and degeneracy operators s_i of $\mathrm{diag}(X)$ are given in terms of those of X by the formulas $d_i = d_i^h \cdot d_i^v$, $s_i = s_i^h \cdot s_i^v$.*

This proposition states that if the simplicial space X is interpreted as a functor

$$X : \Delta^{\mathrm{op}} \times \Delta^{\mathrm{op}} \to \mathbf{Sets}\,,$$

then the realization of X is isomorphic to the functor $\Delta^{\mathrm{op}} \to \mathbf{Sets}$ obtained by composing X with the diagonal functor $\Delta^{\mathrm{op}} \to \Delta^{\mathrm{op}} \times \Delta^{\mathrm{op}}$. This is the reason for the notation "$\mathrm{diag}(X)$". In fact, because of 4.6, the realization of a simplicial space X is often called the *diagonal* of X.

4.7. *A homology spectral sequence.* For a simplicial space X there is an increasing filtration

$$F_0 \,\mathrm{diag}(X) \subset \cdots \subset F_n \,\mathrm{diag}(X) \subset \cdots$$

of $\mathrm{diag}(X)$ given by letting $F_n \,\mathrm{diag}(X)$ be the image (4.4) of the space $\cup_{i \leq n} X_i \times \Delta[n]$. This gives increasing filtrations of the simplicial abelian group $\mathbb{Z}[X]$ (3.18) and of the chain complex $N(\mathbb{Z}[X])$. Associated to this filtration is a spectral sequence for $H_*(\mathrm{diag}(X); \mathbb{Z})$. It turns out to be a first quadrant spectral sequence of homological type (e.g. the differential d^r has bidegree $(-r, r-1)$). We will denote this spectral sequence $E_{i,j}^r(X; \mathbb{Z})$; if there is a coefficient module

M the corresponding spectral sequence is written $E^r_{i,j}(X; M)$. The E^2-page is described by the following proposition. Let $H_n(X; M)$ denote the simplicial abelian group obtained by applying the functor $H_n(-; M)$ (cf. 3.24) dimension-wise to X (so that $H_n(X; M)_j = H_n(X_j; M)$) and let $H_m\, H_n(X; M)$ denote the m'th homology group of $H_n(X; M)$ (cf. 3.23).

4.8. Proposition. [6, XII 5.7] *If X is a simplicial space, there are natural iso-morphisms*

$$E^2_{i,j}(X; M) \approx H_i\, H_j(X; M) \ .$$

The diagonal construction for a simplicial space has the following basic homotopy invariance property. It's not hard to give a direct proof by examining how $\mathrm{diag}(X)$ is glued together from the spaces $X_n \times \Delta[n]$.

4.9. Proposition. [6, XII 4.2, 4.3] *If $f : X \to Y$ is a map of simplicial spaces which induces weak equivalences (3.20) $X_i \to Y_i$ ($i \geq 0$), then f induces a weak equivalence $\mathrm{diag}(X) \to \mathrm{diag}(Y)$.*

4.10. Homotopy colimits. Suppose that $\mathbf{D}$ is a small category. Consider the poset $\mathbf{n}$ as a category with one morphism $i \to j$ if $i \leq j$, and no other mor-phisms. The *singular complex* or *nerve* $N(\mathbf{D})$ of $\mathbf{D}$ is the space given by

$$N(\mathbf{D})_n = \mathrm{Hom}_{\mathbf{Cat}}(\mathbf{n}, \mathbf{D}) \ .$$

Here $\mathbf{Cat}$ is the category in which the objects are small categories and the morphisms are functors. More concretely, an n-simplex σ of $N(\mathbf{D})$ is just a length n sequence

(4.11) $\sigma(0) \xrightarrow{\alpha_1} \sigma(1) \xrightarrow{\alpha_2} \cdots \xrightarrow{\alpha_n} \sigma(n)$

of composable arrows in $\mathbf{D}$. The face and degeneracy operators of $N(\mathbf{D})$ are given by composition with the functors determined by the formulas (3.12). In other words, $d_i\sigma$ is obtained from σ by leaving out $\sigma(0)$ if $i = 0$, by composing α_{i+1} with α_i if $0 < i < n$, and by leaving out $\sigma(n)$ if $i = n$; $s_i\sigma$ is obtained by inserting the identity map $\sigma(i) \to \sigma(i)$ between α_i and α_{i+1}.

4.12. *Example.* Any poset S can be treated as a category in which the objects are the elements of S and in which there is exactly one morphism $x \to y$ if $x \leq y$ (there are no other morphisms). The nerve $N(S)$ is then the same (3.19) as the ordered simplicial complex associated to S in 3.6.

Suppose now that $F : \mathbf{D} \to \mathbf{Sp}$ is a functor. (Such an F is sometimes called a *diagram of spaces with the shape of* $\mathbf{D}$.) The *simplicial replacement* of F is the simplicial space $\coprod_* F$ which in dimension n consists of the coproduct (i.e. disjoint union)

$$\textstyle (\coprod_* F)_n = \coprod_{\sigma \in N(\mathbf{D})_n} F(\sigma(0)) \ .$$

The horizontal degeneracy operator $s_i : (\coprod_* F)_n \to (\coprod_* F)_{n+1}$ maps the space $F(\sigma(0))$ indexed by σ to $F((s_i\sigma)(0))$ by the identity map; the horizontal face operator d_i maps $F(\sigma(0))$ to $F((d_i\sigma)(0))$ by the identity map if $i > 0$ and by the map α_0 if $i = 0$.

> The simplicial space defined here as $\coprod_* F$ differs from the one described in [6, XII.5.1] by an automorphism of the category Δ^{op}. This is a technical point which does not affect any of its properties.

4.13. Definition. Let $\mathbf{D}$ be a small category and $F : \mathbf{D} \to \mathbf{Sp}$ a functor. The *homotopy colimit* of F is the space $\mathrm{hocolim}(F)$ given by $\mathrm{diag}(\coprod_* F)$.

4.14. *Remark.* The homotopy colimit construction is functorial, in the sense that a natural transformation $\tau : F \to F'$ induces a map

$$\mathrm{hocolim}\,\tau : \mathrm{hocolim}\,F \to \mathrm{hocolim}\,F' \ .$$

It follows from 4.9 that homotopy colimits have a strong homotopy invariance property: if τ induces a weak equivalence $\tau_d : F(d) \to F'(d)$ for each object d of $\mathbf{D}$, then $\mathrm{hocolim}(\tau)$ is a weak equivalence.

The homotopy colimit construction is also functorial in $\mathbf{D}$, in the sense that if $j : \mathbf{D}' \to \mathbf{D}$ is a functor and $j^*(F)$ denotes the composite $F \cdot j$, then there is a natural map

$$\mathrm{hocolim}\,j^*(F) \to \mathrm{hocolim}\,F \ .$$

4.15. *Remark.* More explicitly, the space $\mathrm{hocolim}(F)$ can be constructed [6, XII, §2] by taking

- for each object d_0 of $\mathbf{D}$ a copy of $F(d_0)$,
- for each arrow $\alpha_0 : d_0 \to d_1$ of $\mathbf{D}$ a copy of $F(d_0) \times \Delta[1]$, and in general,
- for each chain $d_0 \to \cdots \to d_n$ of composable arrows in $\mathbf{D}$ a copy of $F(d_0) \times \Delta[n]$,

and making appropriate identifications. The identifications amount to

- collapsing $F(d_0) \times \Delta[n]$ if it arises from a chain $d_0 \to \cdots \to d_n$ containing an identity map, and
- identifying the subspace $F(d_0) \times \partial\Delta[n]$ of $F(d_0) \times \Delta[n]$ with an appropriate subspace of $\mathrm{hocolim}(F)$ arising from chains of smaller length.

There is a more or less obvious natural map

$$\mathrm{hocolim}\,F \to \mathrm{colim}\,F \ .$$

This map is not usually a weak equivalence [6, XII.2.5].

4.16. *Remark.* It follows from 4.7 and the construction of hocolim(F) that for any abelian group M there is a natural spectral sequence $E^r_{i,j}(F; M)$ converging to $H_*(\text{hocolim } F; M)$. This is called the *Bousfield-Kan homology spectral sequence* of the homotopy colimit. Let **Ab** be the category of abelian groups and $\mathbf{Ab^D}$ the category of functors $\mathbf{D} \to \mathbf{Ab}$ (with morphisms being natural transformations). According to [6, XII.5.7] the E^2-page $E^2_{i,j}(F; M)$ can be described by the formula

$$E^2_{i,j}(F; M) = \text{colim}_i H_j(F; M) .$$

where

- colim is the colimit functor $\mathbf{Ab^D} \to \mathbf{Ab}$,
- colim_i is the i'th left derived functor of colim, and
- $H_j(F; M)$ is the functor $\mathbf{D} \to \mathbf{Ab}$ obtained by composing F with the homology functor $H_j(-; M)$.

What are the projective objects of $\mathbf{Ab^D}$? Show that there are enough of them to construct left derived functors. See [16, p. 154]. There is a parallel cohomology spectral sequence

$$E^{i,j}_2(F; M) = \lim^i H^j(F; M) \Rightarrow H^{i+j}(\text{hocolim } F; M) ,$$

where

- lim is the limit functor $\mathbf{Ab^{D^{op}}} \to \mathbf{Ab}$,
- $\lim^i$ is the i'th right derived functor of lim, and
- $H^j(F; M)$ is the functor $\mathbf{D^{op}} \to \mathbf{Ab}$ obtained by composing F with the cohomology functor $H^j(-; M)$.

This cohomological version appears more frequently in the literature than the homological one. If $M = \mathbb{F}_p$, for instance, this cohomology spectral sequence is the $\mathbb{F}_p$-dual of $E^r_{i,j}(F; M)$.

The left derived functors of colimit are sometimes easier to calculate than you might think. Suppose that $\mathbf{D}$ is the pushout category of 4.18 below. Show that if

$$F = (A \xleftarrow{u} B \xrightarrow{v} C)$$

is a functor from $\mathbf{D}$ to abelian groups, then

$$\text{colim}_i F = \begin{cases} \text{coker } B \xrightarrow{(u,-v)} A \oplus C & i = 0 \\ \ker B \xrightarrow{(u,-v)} A \oplus C & i = 1 \\ 0 & \text{otherwise} \end{cases}$$

Conclude that the homology spectral sequence for homotopy pushouts amounts to the usual Mayer-Vietoris sequence. If $\mathbf{G}$ is the category of a group G (5.9) and M is a G-module, treated as a functor from $\mathbf{G}$ to abelian groups, observe that $\text{colim } M = \text{colim}_0 M = H_0(G; M)$ (group homology, as in 2.8) and conclude that $\text{colim}_i M$ is isomorphic to $H_i(G; M)$.

Examples of homotopy colimits. How complicated a particular homotopy colimit construction turns out to be depends mostly on the shape of the indexing category $\mathbf{D}$. In the following examples, we assume that F is a functor from $\mathbf{D}$ to $\mathbf{Sp}$.

4.17. *Homotopy coproducts.* If $\mathbf{D}$ is a *trivial category* with a collection $\{d_\alpha\}$ of objects and no nonidentity morphisms, then $\mathrm{hocolim}(F)$ is $\coprod_\alpha F(d_\alpha)$ and the map $\mathrm{hocolim}(F) \to \mathrm{colim}(F)$ is an isomorphism. For example, let $\mathbf{D}$ be an arbitrary category, let d be an object of $\mathbf{D}$, and let $i_d : \{d\} \to \mathbf{D}$ be the inclusion functor whose domain is the trivial category with the single object d. By naturality (4.14) there is an induced map

$$j_d : F(d) = \mathrm{hocolim}\, i_d^*(F) \to \mathrm{hocolim}\, F \ .$$

This shows that each space which appears as a value of the functor F has a natural map to $\mathrm{hocolim}\, F$.

4.18. *Homotopy pushouts.* Let 0 and 1 denote the two copies of $\Delta[0]$ inside the simplicial interval $\Delta[1]$. If $\mathbf{D}$ is the category

$$a \xleftarrow{\ f\ } b \xrightarrow{\ g\ } c$$

then $\mathrm{hocolim}(F)$ is isomorphic to the space obtained from the coproduct

$$F(a) \ \coprod\ (F(b)_f \times \Delta[1]) \ \coprod\ F(b) \ \coprod\ (F(b)_g \times \Delta[1]) \ \coprod\ F(c)$$

by gluing

- $F(b)_f \times 1$ to $F(a)$ by $F(f)$,
- $F(b)_f \times 0$ to $F(b)$ by the identity map of $F(b)$,
- $F(b)_g \times 0$ to $F(b)$ by the identity map of $F(b)$, and
- $F(b)_g \times 1$ to $F(c)$ by $F(g)$.

This is a slightly modified form of the usual double mapping cone construction of the homotopy pushout. The map from $\mathrm{hocolim}(F)$ to $\mathrm{colim}(F)$ is not usually a weak equivalence.

Sequential colimits. If $\mathbf{D}$ is the category given by the poset of nonnegative integers

$$1 \to 2 \to \cdots \to n \to \cdots$$

then, although $\mathrm{hocolim}(F)$ on the face of it looks large and complicated, the map $\mathrm{hocolim}(F) \to \mathrm{colim}(F)$ is always a weak equivalence. The same result holds if $\mathbf{D}$ is replaced by a more general "right filtering" category [6, XII.3.5].

4.19. *Nerves.* If F is the constant functor $*$ with value the one-point space, then $\mathrm{hocolim}(F)$ is isomorphic to the nerve $\mathrm{N}(\mathbf{D})$. More generally, if F is the constant functor with value X, then $\mathrm{hocolim}\, F$ is isomorphic to $X \times \mathrm{N}(\mathbf{D})$. For any F, the unique natural transformation $F \to *$ induces a map $\mathrm{hocolim}\, F \to \mathrm{N}(\mathbf{D})$.

In general, the map q : hocolim $F \to \mathrm{N}(\mathbf{D})$ is a kind of "singular fibration" in which, after geometric realization, the fibre over each point of the interior of a nondegenerate simplex σ is homeomorphic to $|F(\sigma(0))|$. One form of Quillen's Theorem B [31, p. 97] is the statement that if F carries each morphism of $\mathbf{D}$ to a weak equivalence, then $|q|$ is up to homotopy a fibration in which the fibre over the point corresponding to an object d of $\mathbf{D}$ is weakly equivalent to $|F(d)|$.

Simplicial objects. [6, XII.3.4] If $\mathbf{D} = \Delta^{\mathrm{op}}$, so that F is a simplicial space, then hocolim(F) is weakly equivalent to diag(F). In particular, any space X is weakly equivalent to hocolim(F), where $F : \Delta^{\mathrm{op}} \to \mathbf{Sets}$ has $F(\mathbf{n}) = X_n$ (see 3.19 for how to treat a set as a space).

We will see other examples of homotopy colimits later on.

5. Nerves of categories and the Grothendieck construction

The nerve of a category (4.10) is a particularly simple homotopy colimit (4.19). In this section we will describe some special properties of nerves, and then give a construction (due to Thomason [35]) which allows many homotopy colimits to be interpreted as the nerves of auxiliary categories.

Properties of the nerve. Suppose that X and Y are spaces (3.18). Two maps $f, g : X \to Y$ are said to be *simplicially homotopic* if there is a map

$$H : X \times \Delta[1] \to Y$$

such that the restriction of H to $X \times 0$ (4.18) is f and the restriction of H to $X \times 1$ is g (see [24, 5.1] for a daunting combinatorial formulation). The notion of simplicial homotopy is unsatisfactory in some ways; in particular, it does *not* give an equivalence relation on the set of maps from X to Y. Nevertheless, since the geometric realization functor preserves products and the realization of $\Delta[1]$ is a unit interval, the following proposition is clear.

5.1. Proposition. *Suppose that f and g are maps of spaces. If f is simplicially homotopic to g, then $|f|$ is homotopic (in the topological sense) to $|g|$. In particular, f is a weak equivalence (3.20) if and only if g is.*

In practice, explicit simplicial homotopies arise most frequently by applying the nerve construction to natural transformations between functors.

5.2. Proposition. *Suppose that F and G are functors between small categories. If F and G are related by a natural transformation, then $\mathrm{N}(F)$ is simplicially homotopic to $\mathrm{N}(G)$. In particular, $\mathrm{N}(F)$ is a weak equivalence if and only if $\mathrm{N}(G)$ is.*

Proof. The fact that F and G are related by a natural transformation is equivalent to the statement that F and G extend to a functor

$$\tau : \mathbf{D} \times \mathbf{1} \to \mathbf{D}' .$$

(Here $\mathbf{1}$ is the category of the poset $\mathbf{1}$ (4.10)). By construction the functor $N(-) : \mathbf{Cat} \to \mathbf{Sp}$ preserves products, and $N(\mathbf{1})$ is exactly $\Delta[1]$, so $N(\tau)$ gives the required simplicial homotopy. $\qquad\square$

Proposition 5.2 has some immediate consequences.

5.3. Proposition. *If $F : \mathbf{D} \to \mathbf{D}'$ is an equivalence of categories, then $N(F) : N(\mathbf{D}) \to N(\mathbf{D})'$ is a weak equivalence of spaces.*

Proof. If $F' : \mathbf{D}' \to \mathbf{D}$ is an inverse equivalence, then the composites FF' and $F'F$ are naturally equivalent to the respective identity functors. $\qquad\square$

> The conclusion of 5.3 continues to hold under the weaker assumption that there is a functor $F' : \mathbf{D}' \to \mathbf{D}$ which is either left adjoint [22, p .81] or right adjoint to F. In either case, the composites FF' and $F'F$ are connected to the respective identity functors by natural transformations. These natural transformations are not necessarily natural equivalences, but they still provide simplicial homotopies. (5.2).

An object d of a category $\mathbf{D}$ is said to be an *initial object* (resp. a *terminal object*) if for any object d' of $\mathbf{D}$ there is exactly one morphism $d \to d'$ (resp. $d' \to d$). For instance, the empty topological space is an initial object of $\mathbf{Top}$, and any one-point space is a terminal object.

5.4. Proposition. *If $\mathbf{D}$ has either an initial object or a terminal object, then $N(\mathbf{D})$ is weakly contractible.*

Proof. Let $*$ denote the trivial category with one object and only the identity morphism. There is a unique functor $F : \mathbf{D} \to *$, as well as a unique functor $F' : * \to \mathbf{D}$ sending the object of $*$ to d. The composite FF' is the identity functor of $*$. If d is either an initial or a terminal object of $\mathbf{D}$, there is an obvious natural transformation connecting the identity functor of $\mathbf{D}$ to the composite $F'F$. By 5.2, $N(\mathbf{D})$ is weakly equivalent to $N(*) = \Delta[0]$. $\qquad\square$

5.5. *Example.* If S is a poset with a maximal or minimal element, then the nerve of S (4.12) is weakly contractible.

An argument similar to the previous one gives the following.

5.6. Proposition. *If the identity functor of $\mathbf{D}$ is connected to some constant functor $\mathbf{D} \to \mathbf{D}$ by zigzag of natural transformations, then $N(\mathbf{D})$ is weakly contractible.*

A *constant functor* is one which takes the same value d_0 on each object of $\mathbf{D}$ and sends all morphisms of $\mathbf{D}$ to the identity map of d_0. The hypothesis of the proposition means that there exists some chain of natural transformations

$$F_0 \leftarrow F_1 \rightarrow F_2 \leftarrow \cdots \rightarrow F_n$$

in which F_0 is the identity functor and F_n is a constant functor.

The following result is a bit trickier to prove with category theoretic arguments.

5.7. Proposition. [31, p. 86] *The nerve of $\mathbf{D}$ is weakly equivalent in a natural way to the nerve of $\mathbf{D}^{op}$*

5.8. *Remark.* It is pretty clear that there is in general no functor $\mathbf{D} \rightarrow \mathbf{D}^{op}$ inducing a weak equivalence of nerves. Quillen's proof of 5.7 proceeds by introducing a third category $S(\mathbf{D})$ and two functors $s : S(\mathbf{D}) \rightarrow \mathbf{D}^{op}$, $t : S(\mathbf{D}) \rightarrow \mathbf{D}$, both of which do induce weak equivalences when the nerve construction is applied. All of this data depends functorially on $\mathbf{D}$; this is the meaning of the phrase "in a natural way" in 5.7. Another proof of 5.7 is to notice that the geometric realization of $N(\mathbf{D})$ is naturally homeomorphic to the geometric realization of $N(\mathbf{D})^{op}$.

We can now give a few more examples of nerves.

5.9. *Classifying spaces as nerves.* Let G be a discrete group. The *category of G* is the category $\mathbf{G}$ with one object $*$, and with the monoid of maps $* \rightarrow *$ isomorphic to G. A functor $F : \mathbf{G} \rightarrow \mathbf{Sp}$ is essentially a space $X = F(*)$ with an action (3.26) of G. We claim that the nerve $N(\mathbf{G})$ is a classifying space BG. To see this, let $\mathbf{EG}$ denote the category whose objects are the elements x of G, and in which there is exactly one morphism between any two objects, and let $EG = N(\mathbf{EG})$. The space EG is weakly contractible (5.4), because every object of $\mathbf{EG}$ is both initial and terminal. The group G acts on $\mathbf{EG}$ (via functors) by the rule $g \cdot x = gx$. The induced action of G on EG is free, and it is easy to see that the quotient $(EG)/G$ is isomorphic to $N(\mathbf{G})$. Therefore (3.26), $N(\mathbf{G})$ is a classifying space for G. From now on we will let BG denote this specific classifying space. Note that BG is functorial in G, in the sense that a homomorphism $G \rightarrow H$ induces a functor $\mathbf{G} \rightarrow \mathbf{H}$ and hence a map $BG \rightarrow BH$.

5.10. *Groupoids.* A *groupoid* is a small category in which every morphism is invertible. Suppose that $\mathbf{D}$ is a groupoid. For any object x of $\mathbf{D}$ the *vertex group* G_x is the group $\mathrm{Hom}_{\mathbf{D}}(x, x)$. There is an obvious functor $\mathbf{G}_x \rightarrow \mathbf{D}$ taking the unique object of $\mathbf{G}_x$ to x. Choose representatives $\{x_\alpha\}$ of isomorphism classes of objects from $\mathbf{D}$, and consider the functor

$$\coprod_\alpha \mathbf{G}_{x_\alpha} \rightarrow \mathbf{D} \ .$$

It is not hard to see that this functor is an equivalence of categories [22, p. 91] and so induces a weak equivalence on nerves. This gives a weak equivalence

$$\coprod_{\alpha} BG_{x_\alpha} \to N(\mathbf{D}) \ .$$

In other words, the nerve of a groupoid is a disjoint union of classifying spaces of groups; there is one component for each isomorphism class of objects in the groupoid, and this component has the homotopy type of the classifying space of the group of automorphisms of any object in the isomorphism class.

In calculating with nerves we will use the following observation.

5.11. Proposition. *Suppose that G is a group acting on a category $\mathbf{D}$, and that $\mathbf{D}^G$ is the fixed subcategory (that is, the subcategory of $\mathbf{D}$ containing those objects and morphisms fixed by the action). Then the natural map*

$$N(\mathbf{D}^G) \to (N(\mathbf{D}))^G$$

is an isomorphism.

The Grothendieck construction. Suppose that $F : \mathbf{D} \to \mathbf{Sets}$ is a functor. The *transport category* $\mathrm{Tr}(F)$ of F is the category whose objects consist of pairs (d, x), where d is an object of $\mathbf{D}$ and $x \in F(d)$. A map $(d, x) \to (d', x')$ is a morphism $f : d \to d'$ in $\mathbf{D}$ such that $F(f)(x) = x'$. These morphisms compose according to the composition of morphisms in $\mathbf{D}$. The next proposition points out that this construction gives a categorical model for $\mathrm{hocolim}\, F$.

5.12. Proposition. *For any functor $F : \mathbf{D} \to \mathbf{Sets}$ there is a natural isomorphism*

$$N(\mathrm{Tr}(F)) \approx \mathrm{hocolim}\, F \ .$$

Proof. Check that an n-simplex of $\mathrm{hocolim}\, F$ amounts to a pair (σ, x), where σ is an n-simplex (4.11) of the nerve of $\mathbf{D}$ and $x \in F(\sigma(0))$. These are exactly the n-simplices of $N(\mathrm{Tr}(F))$. $\square$

5.13. *Example.* Let X be a space, considered as a functor $X : \Delta^{\mathrm{op}} \to \mathbf{Sets}$. The transport category $\mathrm{Tr}(X)$ can be thought of as a category whose objects are the simplices x of X; a morphism $x \to x'$ is a suitable morphism Φ of Δ^{op} (i.e. a composite of face and degeneracy operators (3.14)) such that $\Phi(x) = x'$. The nerve of $\mathrm{Tr}(X)$ is isomorphic to $\mathrm{hocolim}\, X$, and so is weakly equivalent to X (4). This shows that every space can be constructed up to weak equivalence as the nerve of a category.

For related results see [25] (every connected space is weakly equivalent to the nerve of a category with a single object, i.e. to the classifying space of a monoid) and [21] (every connected space is homology equivalent to the classifying space of a discrete group).

5.14. *Example.* Let G be a group and S a G-set, considered as a functor S : $\mathbf{G} \to \mathbf{Sets}$ (5.9). The transport category of S is the groupoid (5.10) whose objects are the elements of S and in which a morphism $s \to s'$ is an element of $g \in G$ such $gs = s'$. The isomorphism classes of objects in this category correspond to the orbits of the action of G on X, and the vertex group of an object $x \in S$ is the isotropy subgroup G_x. By 5.10, $\mathrm{Tr}(S)$ (or hocolim S) is equivalent to a disjoint union $\coprod_\alpha BG_{x_\alpha}$, where $\{x_\alpha\}$ runs through a set of orbit representatives for the action of G on S.

Proposition 5.12 has a wonderful generalization, due to Thomason. Suppose that $\mathbf{D}$ is a small category, and that $F : \mathbf{D} \to \mathbf{Cat}$ is a functor. The *Grothendieck Construction on* F, denoted $\mathrm{Gr}(F)$, is the category whose objects are the pairs (d, x) where d is an object of $\mathbf{D}$ and x is an object of $F(d)$. An arrow $(d, x) \to (d', x')$ in $\mathrm{Gr}(F)$ is a pair (f, g), where $f : d \to d'$ is a morphism in $\mathbf{D}$ and $g : (F(f))(x) \to x'$ is a morphism in $F(d')$. Arrows compose according to the rule $(f, g) \cdot (f', g') = (f'', g'')$, where f'' is the composite $f \cdot f'$ and g'' is the composite of g with the image of g' under the functor $F(f)$.

5.15. Theorem. [35, 1.2] *Suppose that* $\mathbf{D}$ *is a small category and* $F : \mathbf{D} \to \mathbf{Cat}$ *is a functor. Let* $\mathrm{Gr}(F)$ *be the Grothendieck Construction on* F. *Then there is a natural weak equivalence*

$$N(\mathrm{Gr}(F)) \sim \mathrm{hocolim}\, N(F) .$$

Variations. Suppose that $F : \mathbf{D} \to \mathbf{Cat}$ is a functor. Let F^{op} denote the composite of F with the "opposite" construction $\mathbf{Cat} \to \mathbf{Cat}$; note that F^{op} is again a functor $\mathbf{D} \to \mathbf{Cat}$. It follows from 5.15, 5.7, and the homotopy invariance of homotopy colimits (4.14) that the four categories $\mathrm{Gr}(F)$, $\mathrm{Gr}(F)^{\mathrm{op}}$, $\mathrm{Gr}(F^{\mathrm{op}})$ and $\mathrm{Gr}(F^{\mathrm{op}})^{\mathrm{op}}$ *all* have nerves which are weakly equivalent in a natural way to hocolim $N(F)$.

6. Homotopy orbit spaces

In this section we give a description of the *homotopy orbit space*, which is a particular kind of homotopy colimit that plays a key role in the description of homology decompositions.

Suppose that X is a G-space (i.e. a space with an action of the group G), and denote by the same letter X the corresponding (5.9) functor $\mathbf{G} \to \mathbf{Sp}$.

6.1. Definition. The *homotopy orbit space* of the action of G on X, denoted $X_{\mathrm{h}G}$, is the space

$$X_{\mathrm{h}G} = \mathrm{hocolim}_{\mathbf{G}} X .$$

6.2. *Remark.* It is easy to check that the space X_{hG} is isomorphic to the quotient space

$$(X \times EG)/G\,,$$

where EG is the free contractible G-space of 5.9 and G acts diagonally on the product $X \times EG$. In particular there is (essentially) a fibration sequence

(6.3) $X \to X_{hG} \to (EG)/G = BG\,.$

(See [24, II.7] and [28] for information about fibrations of simplicial sets). The map $X_{hG} \to (*)_{hG} = BG$ is induced by the unique G-map $X \to *$. The space X_{hG} is sometimes called *the fibration over BG associated to the action of G on X* or the *Borel construction* of the action of G on X. Note that $\mathrm{colim}_G X$ is the orbit space of this action, so that there is a natural map (4.15)

$$X_{hG} \to X/G$$

from the homotopy orbit space to the usual orbit space. The property which distinguishes the homotopy orbit space from the usual orbit space is that (like the homotopy pushout) the homotopy orbit space is homotopy invariant (4.14): if $f : X \to Y$ is a map of G-spaces which is an ordinary weak equivalence of spaces, then $(f)_{hG}$ is a weak equivalence.

The homology spectral sequence (4.16) of X_{hG} can be identified with the Serre spectral sequence of 6.3; it has the form

$$E^2_{i,j} = \mathrm{H}_i(BG; \mathrm{H}_j(X; M)) \Rightarrow \mathrm{H}_{i+j}(X_{hG}; M)\,.$$

6.4. *Example.* If S is a transitive G-set, then by 5.14 the space S_{hG} is weakly equivalent to BG_s, where G_s is the isotropy subgroup of some element $s \in S$.

Let $\mathbf{GSp}$ denote the category of G-spaces. If $\mathbf{D}$ is a small category and $F : \mathbf{D} \to \mathbf{GSp}$ is a functor, then, either by inspection or by general naturality principles, $\mathrm{hocolim}(F)$ is in a natural way a G-space. The following proposition could be rephrased as a statement that homotopy limits commute.

6.5. Proposition. *Suppose that G is a group, $\mathbf{D}$ is a small category, and $F : \mathbf{D} \to \mathbf{GSp}$ is a functor. Then there is a natural isomorphism of spaces*

$$(\mathrm{hocolim}\, F)_{hG} \approx \mathrm{hocolim}(F_{hG})\,.$$

Proof. Check using 4.6 that the n-simplices of both spaces correspond to triples (x, σ, τ) where τ is an n-simplex of $N(\mathbf{G})$ (5.9), σ is an n-simplex of $N(\mathbf{D})$, and x is an n-simplex of $\sigma(0)$.   □

7. Homology decompositions

Suppose that G is a discrete group. Recall (2.9) that p is a fixed prime number.

7.1. Definition. A *homology decomposition* for BG consists of a mod p homology isomorphism

$$\text{hocolim } F \xrightarrow[p]{\sim} BG$$

where $\mathbf{D}$ is a small category, $F : \mathbf{D} \to \mathbf{Sp}$ is a functor, and, for each object d of $\mathbf{D}$, $F(d)$ is weakly equivalent to BH_d for some subgroup H_d of G.

7.2. Remark. In any reasonable homology decomposition, the composite map

$$BH_d \simeq F(d) \to \text{hocolim } F \ \to BG$$

(see 4.17) agrees in some homotopical sense with the map $BH_d \to BG$ induced (5.9) by the inclusion of H_d in G as a subgroup.

7.3. Remark. Finding a good homology decomposition is a process of striking a balance. On one extreme, if $\mathbf{D}$ is the trivial category with one object and F is the constant functor with value BG, there is a weak equivalence (even an isomorphism)

$$\text{hocolim } F \approx BG \ .$$

This is a homology decomposition in which all of the complexity is concentrated in the functor F. On the other extreme, if $\mathbf{D}$ is category of the group G (or the transport category (5.13) of BG) and F is the constant functor with value the one-point space ($= B\{e\}$), there is also a weak equivalence hocolim $F \simeq$ BG . In this case F is trivial, and the complexity of the decomposition is concentrated in the shape of the category $\mathbf{D}$. Neither of these decompositions is very interesting. In the most useful homology decompositions the category $\mathbf{D}$ is simple, the subgroups H_d are small, and the focus of the structure is p-primary, in the sense that the map hocolim $F \to BG$ is a mod p homology isomorphism but *not* a weak equivalence.

We will get all of our homology decompositions by the method described in the following proposition.

7.4. Proposition. *Suppose that $\mathbf{D}$ is a small category and that F is a functor from $\mathbf{D}$ to the category of transitive G-sets. Suppose that the natural map (6.2)*

(7.5) $(\text{hocolim } F)_{hG} \to BG$

is a mod p homology isomorphism. Then there is a homology decomposition

$$\text{hocolim}(F_{hG}) \xrightarrow[p]{\sim} BG \ .$$

Proof. It is easy to see that the natural maps $F(d)_{hG} \to BG$ induce a map hocolim$(F_{hG}) \to BG$ which under the equivalence of 6.5 corresponds to the mod p homology isomorphism (7.5). In order to finish the proof, then, all we have to do is check that each space $F(d)_{hG}$ is weakly equivalent to BH_d for some subgroup H_d of G. This follows from 6.4.   □

Any homology decomposition can be constructed as in 7.4. Can you prove this?

Proposition 7.4 suggests that to construct a homology decomposition for BG, the first thing to search for is a G-space X with the property that the natural map $X_{hG} \to BG$ is a mod p homology equivalence. We will find such spaces by looking at complexes associated to *collections* of subgroups of G.

7.6. Definition. A *collection* C of subgroups of G is a set of subgroups of G which is closed under conjugation, in the sense that if $H \in C$ and $g \in G$, then $gHg^{-1} \in C$.

A collection C of subgroups is a poset with respect to inclusion of one subgroup in another, i.e., with respect to the convention that $H \leq H'$ if $H \subset H'$. Let K_C be the associated ordered simplicial complex (3.6), or equivalently the nerve of the category (4.12) associated to C. We will let $\mathbf{K}_C$ denote this category, so that $K_C = N(\mathbf{K}_C)$. The n-simplices of K_C are just chains

$$H_0 \subset H_1 \subset \cdots \subset H_n$$

of elements of C; in particular, if G is finite then K_C is a finite simplicial complex. The group G acts on C by conjugation (if $g \in G$ and $H \in C$ then $g \cdot H = gHg^{-1}$). This action preserves the inclusion relation and passes to an action of G on K_C.

7.7. Definition. A collection C of subgroups of G is said to be *ample* if the natural map

(7.8) $$(K_C)_{hG} \to BG$$

is a mod p homology isomorphism.

7.9. Example. If K_C is contractible the collection C is ample, since then the map 7.8 is a weak equivalence (4.14). The homology decomposition maps (7.1) associated below to such a collection are weak equivalences, not just mod p homology isomorphisms.

At this point it is probably not clear that there exist *any* ample collections of subgroups, Before looking for some, we want to motivate the search. Given a collection C, we will associate to it three functors

(7.10)
$$\begin{aligned} \tilde{\alpha}_C : & (\mathbf{A}_C)^{\mathrm{op}} & \to & \quad \mathbf{GSets} \\ \tilde{\beta}_C : & \mathbf{O}_C & \to & \quad \mathbf{GSets} \\ \tilde{\delta}_C : & \bar{s}d\mathbf{K}_C & \to & \quad \mathbf{GSets} \end{aligned}$$

Of course, one issue will be to define the domain categories of these functors. We will then prove the following proposition. (The proof involves combining 5.12 with 5.3. The most interesting feature of it is the way in which working with functors and natural transformations makes it possible to construct what amount to explicit simplicial homotopies (§5) with only a small amount of work.)

7.11. Proposition. *If C is an ample collection of subgroups of G, then all three functors of (7.10) satisfy the hypotheses of 7.4, and the corresponding functors*

$$\alpha_C = (\tilde{\alpha}_C)_{hG}, \qquad \beta_C = (\tilde{\beta}_C)_{hG}, \qquad \delta_C = (\tilde{\delta}_C)_{hG}$$

give homology decomposition for BG. Conversely, if any one of these functors gives a homology decomposition for BG, then C is ample.

The three decompositions of 7.11 are called the *centralizer decomposition,* the *subgroup decomposition,* and the *normalizer decomposition* associated to C. The names come from the fact that

- the values of α_C have the homotopy type of BC, where C is the centralizer in G of some element of C,
- the values of β_C have the homotopy type of BH, where H is an element of C, and
- the values of δ_C have the homotopy type of BN, where N is an intersection in G of normalizers of elements of C.

The centralizer decomposition. The *C-conjugacy category* $\mathbf{A}_C$ is the category in which the objects are pairs (H, Σ), where H is a group and Σ is a conjugacy class of monomorphisms $i : H \to G$ with $i(H) \in C$. A morphism $(H, \Sigma) \to (H', \Sigma')$ is a group homomorphism $j : H \to H'$ which under composition carries Σ' into Σ.

> We are going to want to take homotopy colimits over $\mathbf{A}_C$ (actually over its opposite category) and, given the way we have defined $\mathbf{A}_C$, this is not possible: the category $\mathbf{A}_C$ is not small. This difficulty is not serious, since $\mathbf{A}_C$ is equivalent to a small category. What details have to be worked out to make this precise? Can you find an explicit small subcategory of $\mathbf{A}_C$ which is equivalent to $\mathbf{A}_C$?

There is a functor

$$\tilde{\alpha}_C : (\mathbf{A}_C)^{\mathrm{op}} \to \mathbf{GSets}$$

which assigns to each object (H, Σ) the set Σ itself, on which G acts transitively by conjugation. Here is the first third of 7.11.

7.12. Proposition. *The collection C is ample if and only if the natural map*

$$(\mathrm{hocolim}\, \tilde{\alpha}_C)_{hG} \to BG$$

is a mod p homology isomorphism.

Proof. According to 5.12, the space $X_{\mathcal{C}}^{\alpha} = \text{hocolim } \tilde{\alpha}_{\mathcal{C}}$ is the nerve of a category $\mathbf{X}_{\mathcal{C}}^{\alpha}$ whose objects consist of pairs (H, i), where H is a group and $i : H \to G$ is a monomorphism with $i(H) \in \mathcal{C}$. A map $(H, i) \to (H', i')$ is a homomorphism $j : H \to H'$ such that $i'j = i$. The action of G on $\text{hocolim } \tilde{\alpha}_{\mathcal{C}}$ corresponds to the action of G on $\mathbf{X}_{\mathcal{C}}^{\alpha}$ obtained by letting $g \cdot (H, i) = (H, gig^{-1})$. There is a functor

$$u : \mathbf{X}_{\mathcal{C}}^{\alpha} \to \mathbf{K}_{\mathcal{C}}$$

which sends (H, i) to $i(H)$, a functor which by inspection is G-equivariant. To finish the proof it is enough to show that u induces a weak equivalence from the nerve of $\mathbf{X}_{\mathcal{C}}^{\alpha}$ to $K_{\mathcal{C}}$. In fact, if this is true, then by 6.2 there is a commutative diagram

$$
\begin{array}{ccc}
(\text{hocolim } \tilde{\alpha}_{\mathcal{C}})_{hG} = (X_{\mathcal{C}}^{\alpha})_{hG} & \xrightarrow{\ \simeq\ } & (K_{\mathcal{C}})_{hG} \\
\downarrow & & \downarrow \\
BG & \xrightarrow{\ =\ } & BG
\end{array}
$$

in which by 4.14 the top arrow is a weak equivalence. This implies that the left hand arrow is an isomorphism on mod p homology if and only if the right hand one is.

To see that u has the desired property, use 5.3 and note that u is an equivalence of categories. In fact, there is a functor $u' : \mathbf{K}_{\mathcal{C}} \to \mathbf{X}_{\mathcal{C}}^{\alpha}$ which sends an object $H \in \mathcal{C}$ to the pair (H, i) where i is the inclusion map $H \to G$. The composite uu' is the identity map of $\mathbf{K}_{\mathcal{C}}$, while the composite $u'u$ is connected to the identity map of $\mathbf{X}_{\mathcal{C}}^{\alpha}$ by an obvious natural equivalence. $\square$

7.13. Remark. If $H \subset G$ is a subgroup, let $C_G(H)$ denote the centralizer of H in G. By 6.4, the centralizer decomposition functor $\alpha_{\mathcal{C}} = (\tilde{\alpha}_{\mathcal{C}})_{hG}$ assigns to any object (H, Σ) of $\mathbf{A}_{\mathcal{C}}$ a space which, for any $i \in \Sigma$, is weakly equivalent to $BC_G(i(H))$.

The subgroup decomposition. The $\mathcal{C}$ *orbit category* $\mathbf{O}_{\mathcal{C}}$ is the category whose objects are the G-sets G/H, $H \in \mathcal{C}$, and whose morphisms are G-maps. There is a functor $\tilde{\beta}_{\mathcal{C}} : \mathbf{O}_{\mathcal{C}} \to \mathbf{GSets}$ which assigns to G/H the G-set G/H itself. Here is the second third of 7.11.

7.14. Proposition. *The collection $\mathcal{C}$ is ample if and only if the natural map*

$$(\text{hocolim } \tilde{\beta}_{\mathcal{C}})_{hG} \to BG$$

is a mod p homology isomorphism.

Proof. By 5.12, the space $X_{\mathcal{C}}^{\beta} = \text{hocolim } \tilde{\beta}_{\mathcal{C}}$ is the nerve of a category $\mathbf{X}_{\mathcal{C}}^{\beta}$ whose objects consist of pairs $(G/H, x)$, where G/H is a coset space of G with $H \in \mathcal{C}$, and x is an element of G/H. A morphism $(G/H, x) \to (G/H', x')$ is a G-map

$f : G/H \to G/H'$ such that $f(x) = x'$. The action of G on $X_{\mathcal{C}}^{\beta}$ corresponds to the action of G on $X_{\mathcal{C}}^{\beta}$ obtained by letting $g \cdot (G/H, x) = (G/H, gx)$. There is a functor

$$v : X_{\mathcal{C}}^{\beta} \to K_{\mathcal{C}}$$

which sends $(G/H, x)$ to the isotropy subgroup G_x, and by inspection this functor is G-equivariant. As in the proof of 7.12, it is enough to show that v is an equivalence of categories. This is easy; there is a functor $v' : K_{\mathcal{C}} \to X_{\mathcal{C}}^{\beta}$ which sends $H \in \mathcal{C}$ to the pair $(G/H, eH)$, where $e \in G$ is the identity element. The composite vv' is the identity functor of $K_{\mathcal{C}}$, and the composite $v'v$ is connected to the identity functor of $X_{\mathcal{C}}^{\beta}$ by an obvious natural equivalence. $\square$

7.15. *Remark.* The subgroup decomposition functor $\beta_{\mathcal{C}} = (\tilde{\beta}_{\mathcal{C}})_{hG}$ assigns to the object G/H of $\mathbf{O}_{\mathcal{C}}$ the space $(G/H)_{hG}$, which by 6.4 is weakly equivalent to BH.

The normalizer decomposition. This one is a little trickier. Let $\bar{s}d K_{\mathcal{C}}$ be the category of "orbit simplices" for the action of G on the simplicial complex $K_{\mathcal{C}}$. The objects of $\bar{s}d K_{\mathcal{C}}$ are the orbits $\bar{\sigma}$ of the action of G on the simplices of $K_{\mathcal{C}}$, and there is one morphism $\bar{\sigma} \to \bar{\sigma}'$ if for some simplices $\sigma \in \bar{\sigma}$ and $\sigma' \in \bar{\sigma}'$, σ' is a face of σ. (This might look backwards, but it's necessary to define morphisms this way in order to get the functor $\tilde{\delta}_{\mathcal{C}}$ below.) Note that here we are definitely thinking of $K_{\mathcal{C}}$ as a simplicial complex, and not as a simplicial set; degenerate simplices play no role. An object of $\bar{s}d K_{\mathcal{C}}$ is an equivalence class, under the action of G by conjugation, of chains

(7.16) $$H_0 \subsetneq H_1 \subsetneq \cdots \subsetneq H_n$$

such that each H_i belongs to $\mathcal{C}$.

There is a functor

$$\tilde{\delta}_{\mathcal{C}} : \bar{s}d K_{\mathcal{C}} \to \mathbf{GSets}$$

which assigns to an orbit $\bar{\sigma}$ the transitive G-set provided by $\bar{\sigma}$ itself. Here is the last part of 7.11.

7.17. Proposition. *The collection $\mathcal{C}$ is ample if and only if the natural map*

$$(\text{hocolim } \tilde{\delta}_{\mathcal{C}})_{hG} \to BG$$

is a mod p homology isomorphism.

Proof. By 5.12, the space $X_{\mathcal{C}}^{\delta} = \text{hocolim } \tilde{\delta}_{\mathcal{C}}$ is the nerve of a category $\mathbf{X}_{\mathcal{C}}^{\delta}$ whose objects are the simplices σ of $K_{\mathcal{C}}$. There is one morphism $\sigma \to \sigma'$ if σ' is a face of σ and no other morphisms. The category $\mathbf{X}_{\mathcal{C}}^{\delta}$ is the opposite of the category whose nerve is the barycentric subdivision (3.6) of $K_{\mathcal{C}}$. There are G-equivariant homeomorphisms (5.8, 3.6)

$$|X_{\mathcal{C}}^{\delta}| \cong |\,\mathrm{N}((\mathbf{X}_{\mathcal{C}}^{\delta})^{\mathrm{op}})| \cong |K_{\mathcal{C}}| \ .$$

However, these homeomorphisms are not realized by functors or simplicial maps, so there is some more work to be done. We leave this to the reader: the problem is to devise a way of relating $X_{\mathcal{C}}^{\delta}$ to $K_{\mathcal{C}}$ inside the category $\mathbf{Sp}$, by maps which are G-equivariant and are weak equivalences. □

7.18. *Remark.* If H is a subgroup of G, let $N_G(H)$ denote the normalizer of H in G. By 6.4, the normalizer decomposition functor $\delta_{\mathcal{C}} = (\tilde{\delta}_{\mathcal{C}})_{h\mathrm{G}}$ assigns to the orbit of the simplex (7.16) of $K_{\mathcal{C}}$ a space which has the homotopy type of $B(\cap_i N_G(H_i))$

8. Sharp homology decompositions. Examples

Associated to any homology decomposition for BG (7.1) is a first quadrant mod p homology spectral sequence (4.16)

$$E_{i,j}^2(F) = \text{colim}_i \, \mathrm{H}_j(F) \Rightarrow \mathrm{H}_{i+j}(BG) \ .$$

8.1. Definition. A homology decomposition for BG is *sharp* if its homology spectral sequence collapses onto the vertical axis, in the sense that $E_{i,j}^2 = 0$ for $i > 0$.

The usefulness of a sharp homology decomposition functor F is that it gives an isomorphism

(8.2) $\text{colim } \mathrm{H}_*(F) = \text{colim}_0 \, \mathrm{H}_*(F) \xrightarrow{\approx} \mathrm{H}_* BG \ .$

This is essentially a formula for $\mathrm{H}_* BG$ in terms of the homology groups of subgroups of G.

It turns out that sharp homology decompositions are sometimes easier to recognize than arbitrary ones. One way in which we will show that a collection $\mathcal{C}$ of subgroups is ample (7.7) is to show that one of the three functors $\alpha_{\mathcal{C}}$, $\beta_{\mathcal{C}}$, or $\delta_{\mathcal{C}}$ associated to $\mathcal{C}$ gives a sharp homology decomposition for BG. The other two functors then give homology decompositions too (although these may or may not be sharp).

Examples of homology decompositions. From now on we will assume that G is a *finite* group.

8.3. Definition. An ample collection C of subgroups of G is said to be *central-izer-sharp* (resp. *subgroup-sharp*, *normalizer-sharp*) if the centralizer decomposition (resp. subgroup decomposition, normalizer decomposition) associated to C is sharp.

Here are some examples of homology decompositions.

8.4. *Example:* $C = \{\{e\}\}$. Suppose that C contains only the trivial subgroup of G. Then K_C is a one-point space, so C is ample. Moreover, C is both centralizer-sharp and normalizer-sharp, since both the centralizer and normalizer decomposition diagrams reduce to the trivial diagram with BG as its only constituent. The subgroup decomposition category $\mathbf{O}_C$ has only one object, and the group G is the space of self-maps of this object. The functor β_C assigns to this object the trivial one-point space. By 6.2, the homology spectral sequence of β_C has

$$E_{i,j}^2 = \begin{cases} H_i(G) & j = 0 \\ 0 & j > 0 \end{cases}.$$

This spectral sequence collapses, but onto the wrong axis! The collection C is subgroup-sharp if and only if the reduced mod p homology of BG is trivial.

> If G is a finite group, the reduced mod p homology of BG is trivial if and only if the order of G is prime to p. Can you prove this? (Hint: the hard part is to show that if the mod p (co-)homology is trivial, then the order of G is prime to p. Suppose that the order of G is not prime to p. Take a faithful complex representation ρ of G, and try computing the Chern classes of the restriction of ρ to a cyclic subgroup of G of order p.)

8.5. *Example:* $C = \{G\}$. Suppose that C contains only G itself. Again K_C is has only one point, so C is ample. The collection C is both subgroup-sharp and normalizer-sharp, since both the subgroup and normalizer decomposition diagrams reduce to the trivial diagram with BG as its only constituent. Let Z be the center of G. The category $\mathbf{A}_C$ has one object whose group of self-maps is G/Z and the functor α_C assigns to this object the space BZ. By 6.2, the homology spectral sequence of α_C is the Lyndon-Hochschild-Serre spectral sequence of the group extension $Z \to G \to G/Z$. From this it is possible to prove that C is centralizer-sharp if and only if Z contains a Sylow p-subgroup of G; this is the case if and only if G is the product of an abelian p-group and a group of order prime to p.

> Can you prove these last statements?

8.6. *Other trivial examples.* If C contains either $\{e\}$ or G, then K_C is contractible, because the category associated to C (4.12) has either $\{e\}$ as an initial object or G as a terminal object (see 5.4). In particular, C is ample. The corresponding homology decompositions are in some sense circular, since (as suggested in the above two examples) G itself is somehow encoded in each of the three homotopy colimits associated to $\mathcal{C}$.

> Suppose that $\mathcal{C}$ is the set of conjugates of H, where H is some subgroup of G different from $\{e\}$ and from G itself. Can you think of examples of subgroups H for which $\mathcal{C}$ is ample? What are the corresponding three decomposition functors? What sharpness properties does $\mathcal{C}$ have? Think first about the special case in which H is normal, so that $\mathcal{C} = \{H\}$.

8.7. *Non-identity p-subgroups.* Suppose that p divides the order of G, and let $\mathcal{C}$ be the collection of all non-identity p-subgroups of G. In one form or another it has been known for a long time that $\mathcal{C}$ is ample. We will prove this in §12, and show in addition that $\mathcal{C}$ is centralizer-sharp, subgroup-sharp, and normalizer-sharp; Most of the technical results that go into proving sharpness are due originally to Webb [36] [37]. The colimit formula (8.2) for $H_* BG$ that arises from the associated subgroup decomposition (§7) is essentially the $\mathbb{F}_p$-dual of the formula of Cartan and Eilenberg [8, p. 259] expressing $H^* BG$ as the set of stable elements in the cohomology of a Sylow p-subgroup of G.

8.8. *Elementary abelian p-subgroups.* A finite abelian group is said to be an *elementary abelian p-group* if it is a module over $\mathbb{F}_p$. Suppose that p divides the order of G, and let $\mathcal{C}$ be the collection of all nontrivial elementary abelian p-subgroups of G, and $\mathcal{C}'$ the collection of all nontrivial p-subgroups of G. It is a theorem of Quillen that the natural map $K_{\mathcal{C}} \to K_{\mathcal{C}'}$ is a weak equivalence [2, 6.6.1].

> One proof of Quillen's theorem goes like this. If $F : \mathbf{A} \to \mathbf{B}$ is a functor between small categories, and b is an object of $\mathbf{B}$, let $F \downarrow b$ denote the category whose objects consist of pairs (a, u) where a is an object of $\mathbf{A}$ and $u : F(a) \to b$ is a morphism in $\mathbf{B}$. A map $(a, u) \to (a', u')$ in this category is a morphism $f : a \to a'$ in $\mathbf{A}$ such that $u' = u \cdot F(f)$. Letting b vary gives a functor $F \downarrow - :$ $\mathbf{B} \to \mathbf{Cat}$. Check that there are functors $\mathrm{Gr}(F \downarrow -) \leftrightarrow \mathbf{A}$ with the property that each composite is connected to the identity functor by a natural transformation. Conclude that the nerve of $\mathrm{Gr}(F \downarrow -)$ is weakly equivalent to the nerve of $\mathbf{A}$. Argue further that if each one of the categories $F \downarrow b$ has a contractible nerve, then the nerve of $\mathrm{Gr}(F \downarrow -)$ is equivalent to the nerve of $\mathbf{B}$. Apply this to the case in which F is the inclusion of posets $\mathcal{C} \to \mathcal{C}'$. Use 12.1 to show that for each object b of $\mathcal{C}'$, the identity functor of $F \downarrow b$ is connected to a constant functor by a zigzag of natural transformations.

Since $\mathcal{C}'$ is ample (8.7), so is $\mathcal{C}$. In fact, $\mathcal{C}$ is both centralizer-sharp and normalizer-sharp; by the duality between homology and cohomology, it cannot possibly be subgroup-sharp unless the mod p cohomology of G is detected on elementary abelian p-subgroups.

The mod p cohomology of G is *detected on elementary abelian subgroups* if for every nonzero $x \in \mathrm{H}^* BG$ there is an elementary abelian p-subgroup V of G such that the restriction of x to $\mathrm{H}^* BV$ is nonzero. Can you think of a finite group G such that $\mathrm{H}^* BG$ is *not* detected on elementary abelian p-subgroups? Quillen showed [30] that if G is the group $\mathrm{GL}_n(\mathbb{F}_q)$ (where q is relatively prime to p) or if G is a symmetric group, then the mod p cohomology of BG *is* detected on elementary abelian subgroups.

We will discuss this and related collections in §13. The question of centralizer sharpness was first studied by Jackowski and McClure [19], later by Dwyer and Wilkerson [14] and by Benson [3]. The main ingredient necessary to prove normalizer sharpness was provided by Brown [7, p. 268].

Let $p = 2$, and let G be the group $\mathrm{SL}_3(\mathbb{F}_q)$, where q is odd. Show that up to conjugacy there are only two nontrivial elementary abelian p-subgroups of G. What are the corresponding homology decompositions?

8.9. *p-centric subgroups.* A p-subgroup P of G is said to be *p-centric* if the center of P is a Sylow p-subgroup of $C_G(P)$. This is equivalent to the condition that $C_G(P)$ is the direct product of the center of P and a group of order prime to p. Let $\mathcal{C}$ be the collection of all p-centric subgroups of G. Then $\mathcal{C}$ is ample [11] as well as subgroup-sharp [12].

Suppose that G is a group with the property that the Sylow p-subgroups of G are abelian. Argue that up to conjugacy there is only one p-centric subgroup of G. What are the associated homology decompositions? Is it obvious that the subgroup decomposition is sharp? Recover a certain classical theorem of Swan.

8.10. *Subgroups both p-centric and p-stubborn.* A p-subgroup P of G is said to be *p-stubborn* or *p-radical* if $N_G(P)/P$ has no non-identity normal p-subgroups. Let $\mathcal{C}$ be the collection of all subgroups of G which are both p-centric (8.9) and p-stubborn. Then $\mathcal{C}$ is ample [11] and subgroup-sharp [12]. The role of p-stubborn subgroups was pointed out by Jackowski, McClure and Oliver [20], and independently (in a different context and with different terminology) by Bouc [2, 6.6.6].

The recent paper of Grodal [18] contains a lot of very interesting additional information about homology decompositions.

9. Reinterpreting the homotopy colimit spectral sequence

In this section we point out that the Bousfield-Kan homology spectral sequence (4.16) associated to a homotopy colimit can be interpreted as a Leray spectral sequence. For the particular homotopy colimits that give rise to the homology decompositions of §7, these Leray spectral sequences are the isotropy spectral sequences associated to G-spaces. This makes it possible to study the spectral sequences (e.g., in order to show that a decomposition is sharp) by making geometric manipulations with G-spaces.

We first introduce the Leray spectral sequence, then we specialize to the isotropy spectral sequence, and finally we explain the relationship of the isotropy spectral sequence to homology decompositions.

The Leray spectral sequence. The n-*skeleton* $\mathrm{sk}_n B$ $(n \geq 0)$ of a space B is the subobject of B generated by all simplices of dimension $\leq n$.

9.1. Definition. Let $f : X \to B$ be a map of spaces, and let X_n denote $f^{-1}(\mathrm{sk}_n B)$. The *Leray spectral sequence* of f is the (mod p) homology spectral sequence associated to the filtration

$$X_0 \subset X_1 \subset \cdots \subset X_n \subset \cdots$$

of X.

This spectral sequence is usually indexed in such a way that $E^1_{i,j} = H_{i+j}(X_i, X_{i-1})$. Since X_n contains $\mathrm{sk}_n X$, $E^1_{i,j} = 0$ for $j < 0$ and this is a first quadrant, strongly convergent homology spectral sequence. In particular, the differential d^r has bidegree $(-r, r-1)$. Here are a few examples.

9.2. *Collapse map.* If Y is a subspace of X, the E^2-term of the Leray spectral sequence of $f : X \to X/Y$ vanishes except for the groups $E^2_{0,0} = H_0 X$, $E^2_{0,j} = H_j(Y)$ $(j > 0)$ and $E^2_{i,0} = H_i(X/Y)$ $(i > 0)$. The various differentials running from the horizontal axis to the vertical axis give the connecting homomorphisms in the long exact homology sequence of the pair (X, Y).

9.3. *Fibration.* If $f : X \to Y$ is a fibration of spaces [24], the Leray spectral sequence of f can be identified with the Serre spectral sequence [28] of $|f|$.

9.4. *Homotopy colimit.* Suppose that $\mathbf{D}$ is a small category and $F : \mathbf{D} \to \mathbf{Sp}$ is a functor. The unique natural transformation from F to the constant functor $*$ with value the one-point space induces a map

$$f : \mathrm{hocolim}\, F \to \mathrm{hocolim}\, * \approx \mathrm{N}(\mathbf{D})$$

The Leray spectral sequence of f can be identified from E^2 onwards with the Bousfield-Kan homology spectral sequence (4.16) of the homotopy colimit. This can be seen by inspecting the definitions (4.7, 9.1) and seeing that the two spectral sequences arise from the same filtration of $\mathrm{hocolim}\, F$.

The isotropy spectral sequence. We will be interested in a specific Leray spectral sequence associated to a G-space X.

9.5. Definition. The *isotropy spectral sequence* of a G-space X is the Leray spectral sequence of the map (6.2)

$$f : X_{\mathrm{hG}} = (EG \times X)/G \to X/G .$$

9.6. *Remark.* Consider the diagram

(9.7)

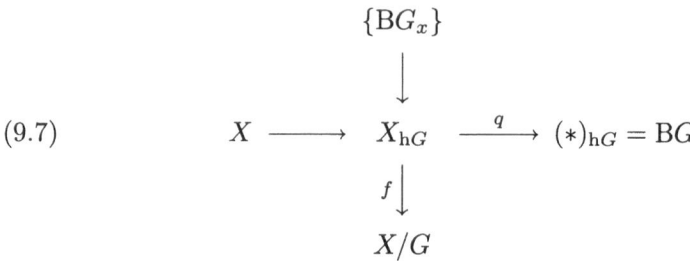

The horizontal sequence is a fibration sequence, and the Leray spectral sequence of q is the usual Serre spectral sequence converging to $H_*(X_{hG}; M)$. Equivalently, this is the Bousfield-Kan homology spectral sequence associated to the description of X_{hG} as a homotopy colimit (6.1). The vertical sequence is meant to suggest that the geometric fibres of f in general differ from point to point, and that the fibre over a simplex $\bar{x} \in X/G$ can be identified up to homotopy with BG_x, where G_x is the isotropy subgroup of a simplex $x \in X$ above $\bar{x}$. The Leray spectral sequence of f is the isotropy spectral sequence of X, and it too converges to $H_*(X_{hG})$.

9.8. *Examples.* If G acts freely on X then the map f of (9.7) is an equivalence with contractible fibres, and the isotropy spectral sequence of X collapses onto the horizontal axis. More generally, suppose that K is a normal subgroup of G which acts trivially on X, and that the quotient group $W = G/K$ acts freely on X. Then the map f of 9.7 is a fibration with fibre BK, and the isotropy spectral sequence of X can be identified with the Serre spectral sequence of the fibration $BK \to X_{hG} \to X_{hW}$.

9.9. *The E^2-page of the isotropy spectral sequence.* The isotropy spectral sequence of the G-space X can be constructed as the homology spectral sequence (4.7) of the simplicial space Y given by

$$Y_n = (X_n)_{hG} \ .$$

Here we are treating X_n as a discrete space (3.19) with a G-action; the horizontal (4.5) face and degeneracy operators are induced by the face and degeneracy operators of X.

According to 4.8 and 3.23, the E^2-page of the isotropy spectral sequence can be identified as follows. Write $H_j(X_i)_{hG}$ for $H_j((X_i)_{hG})$. For each $j \geq 0$ one can form a chain complex $\langle H_j(X_*)_{hG}, d \rangle$ which in dimension $i \geq 0$ contains the group $H_j(X_i)_{hG}$ and has boundary map d induced by taking the alternating sum of the maps

$$H_j(X_i)_{hG} \to H_j(X_{i-1})_{hG}$$

induced by the $(i+1)$ face operators $X_i \to X_{i-1}$. The group in position $E^2_{i,j}$-term of the isotropy spectral sequence is then the i'th homology group of $\langle H_j(X_*)_{hG}, d \rangle$.

Homology decomposition spectral sequences. We now show that the Bousfield-Kan homology spectral sequence of any homotopy colimit constructed by the technique of 7.4 can be interpreted as the isotropy spectral sequence of an associated G-space. This will allow us to work with the Bousfield-Kan spectral sequence (which is pretty abstract) by manipulating G-spaces (which are much easier to handle). The basis for this is the following proposition.

9.10. Proposition. *Suppose that* $\mathbf{D}$ *is a small category and that* F *is a functor from* $\mathbf{D}$ *to the category of transitive G-sets. Let* $X(F)$ *denote the G-space* hocolim F. *Then the Bousfield-Kan homology spectral sequence for the homology of* hocolim(F_{hG}) *can be identified in a natural way with the isotropy spectral sequence of* $X(F)$.

Proof. There is a commutative diagram

$$\begin{array}{ccc} \text{hocolim}(F_{hG}) & \longrightarrow & \text{hocolim}(F/G) \approx \mathrm{N}(\mathbf{D}) \\ \approx \downarrow & & \approx \downarrow \\ (\text{hocolim } F)_{hG} & \longrightarrow & (\text{hocolim } F)/G \end{array}$$

in which both vertical arrows are isomorphisms (see 6.5 for the left arrow). The Bousfield-Kan homology spectral sequence of F is the Leray spectral sequence of the upper map (9.4) and the isotropy spectral sequence of $X(F)$ is the Leray spectral sequence of the lower one. $\qquad \square$

We now establish notation for the G-spaces associated to the decompositions of §7.

9.11. The centralizer decomposition. Let $X^\alpha_{\mathcal{C}}$ denote the G-space given by hocolim $\tilde{\alpha}_{\mathcal{C}}$. As in the proof of 7.12, $X^\alpha_{\mathcal{C}}$ is the nerve of the category $\mathbf{X}^\alpha_{\mathcal{C}}$ whose objects are pairs (H, i), where H is a group and $i : H \to G$ is a monomorphism with $i(H) \in \mathcal{C}$. The Bousfield-Kan homology spectral sequence associated to $\alpha_{\mathcal{C}}$ is the isotropy spectral sequence of the action of G on $X^\alpha_{\mathcal{C}}$.

9.12. The subgroup decomposition. Let $X^\beta_{\mathcal{C}}$ denote the G-space given by hocolim $\tilde{\beta}_{\mathcal{C}}$. As in the proof of 7.14, this is the nerve of a category $\mathbf{X}^\beta_{\mathcal{C}}$ whose objects consist of pairs $(G/H, x)$ where $H \in \mathcal{C}$ and $x \in G/H$. The Bousfield-Kan homology spectral sequence associated to $\beta_{\mathcal{C}}$ is the isotropy spectral sequence of the action of G on $X^\beta_{\mathcal{C}}$.

9.13. The normalizer decomposition. Let $X^\delta_{\mathcal{C}}$ denote the space $K_{\mathcal{C}}$. This is the nerve of the category $\mathbf{X}^\delta_{\mathcal{C}}$ associated to the poset $\mathcal{C}$. It is possible to prove that

X_C^δ is weakly G-equivalent (10.5) to the space hocolim $\tilde{\delta}_C$ (see the remarks in the proof of 7.17), and so the Bousfield-Kan homology spectral sequence associated to δ_C is the isotropy spectral sequence of the action of G on X_C^δ.

10. Bredon homology and the transfer

The previous section identified the Bousfield-Kan spectral sequence derived from a homology decomposition of BG as the isotropy spectral sequence of an associated G-space X. The E^2-term of this isotropy spectral sequence is a type of homological functor of X; in this section we point out exactly what sort of construction this homological functor is, and describe some of its properties.

10.1. Definition. Let $\mathcal{H}$ be a functor from the category of $\mathbb{F}_p[G]$-modules to the category of $\mathbb{F}_p$ vector spaces. The functor $\mathcal{H}$ is said to be *a coefficient functor* for G if $\mathcal{H}$ preserves arbitrary direct sums. If K is a subgroup of G, the *restriction of $\mathcal{H}$ to K*, denoted $\mathcal{H}|_K$, is the coefficient functor for K given by $\mathcal{H}|_K(A) = \mathcal{H}(\mathbb{F}_p[G] \otimes_{\mathbb{F}_p[K]} A)$.

> The functor $\mathcal{H}$ is said to *preserve arbitrary direct sums* if, for any set $\{M_\alpha\}$ of $\mathbb{F}_p[G]$-modules, the natural map
> $$\oplus_\alpha \mathcal{H}(M_\alpha) \to \mathcal{H}(\oplus_\alpha M_\alpha)$$
> is an isomorphism.

10.2. *Example.* For each $j \geq 0$ there is a coefficient functor $\mathcal{H}_j^G$ given by $\mathcal{H}_j^G(A) = \mathrm{H}_j(G; A)$ (see 2.8). These are the coefficient functors we will interested in. If K is a subgroup of G and $\mathcal{H} = \mathcal{H}_j^G$, then $\mathcal{H}|_K \approx \mathcal{H}_j^K$.

> Prove the final statement; it's a form of Shapiro's lemma [38, 6.3.2].

10.3. Definition. Suppose that $\mathcal{H}$ is a coefficient functor for G and that X is a G-space. Let $(C_*^G(X; \mathcal{H}), d)$ be the chain complex with $C_n^G(X; \mathcal{H}) = \mathcal{H}(\mathbb{F}_p[X_n])$ (see 3.24) and with boundary d induced by the alternating sum of the face maps in X. The *Bredon homology groups* of X with coefficients in $\mathcal{H}$, denoted $\mathrm{H}_*^G(X; \mathcal{H})$, are defined to be the homology groups of $C_*^G(X; \mathcal{H})$.

10.4. *Example.* Suppose that X is a G-space, and let $\{\mathcal{H}_j^G\}$ be the coefficient functors from 10.2. The E^2 page of the isotropy spectral sequence for X is then given by
$$E_{i,j}^2 = \mathrm{H}_i^G(X; \mathcal{H}_j^G) .$$
This follows immediately from 9.10, 9.9, and the observation that for any G-set S, there is a natural isomorphism
$$\mathcal{H}_j^G(\mathbb{F}_p[S]) \approx \mathrm{H}_j(S_{\mathrm{h}G}) .$$

Prove this last statement, by combining additivity of $\mathcal{H}_j^G$ with the final sentence of 10.2. It is possible to use coefficients for Bredon homology more general than the functors allowed above; in fact, any functor from the orbit category of G to abelian groups can be used to construct a Bredon homology theory. For practical purposes, the "coefficient functors" above are the same as the "cohomological Mackey functors" of [39] or [34, p. 1928].

A key property of $\mathrm{H}_*^G(X; \mathcal{H})$. The groups $\mathrm{H}_*^G(X; \mathcal{H})$ have a certain basic invariance property.

10.5. Definition. A map $f : X \to Y$ of G-spaces is said to be *a weak G-equivalence* if $f^H : X^H \to Y^H$ is a weak equivalence for every subgroup H of G.

10.6. *Remark.* If X is a G-space, let $\mathrm{Iso}(X)$ denote the set of subgroups of G which appear as isotropy groups of simplices of X. By [11, 4.1], a map $f : X \to Y$ of G-spaces is a weak G-equivalence if and only if it induces a weak equivalence $f^H : X^H \to Y^H$ for all $H \in \mathrm{Iso}(X) \cup \mathrm{Iso}(Y)$.

10.7. Proposition. *Suppose that $f : X \to Y$ is a a weak G-equivalence and that $\mathcal{H}$ is a coefficient functor (10.1). Then f induces isomorphisms $\mathrm{H}_*^G(X; \mathcal{H}) \cong \mathrm{H}_*^G(Y; \mathcal{H})$.*

A proof of this is sketched in §14.

10.8. The transfer. All of our techniques for dealing with the G-spaces from 9.11, 9.12, and 9.13 are based in one way or another on the transfer. We assume from now on that $\mathcal{H}$ is a coefficient functor (10.1), in practice one of the coefficient functors $\mathcal{H}_j^G$ (10.2).

Suppose that $f : S \to T$ is a map of G-sets. There is an induced map $\mathbb{F}_p[S] \to \mathbb{F}_p[T]$, also denoted f, as well as a map

$$f_* = \mathcal{H}(f) : \mathcal{H}(\mathbb{F}_p[S]) \to \mathcal{H}(\mathbb{F}_p[T]) \,.$$

Say that f is *finite-to-one* if for each $x \in T$ the set $f^{-1}(x)$ is finite. For such an f there is a G-map $\tau(f) : \mathbb{F}_p[T] \to \mathbb{F}_p[S]$, called the *pretransfer*, which sends $x \in T$ to $\sum_{y \in f^{-1}(x)} y$. The induced map

$$\tau_*(f) : \mathcal{H}(\mathbb{F}_p[T]) \to \mathcal{H}(\mathbb{F}_p[S])$$

is the *transfer* associated to f.

10.9. *Example.* Suppose that H and K are subgroups of G with $H \subset K$, and let $\mathcal{H} = \mathcal{H}_j^G$ (10.2). Let $f : G/H \to G/K$ be the projection map. By Shapiro's lemma [38, 6.3.2] there are isomorphisms

$$\mathcal{H}(\mathbb{F}_p[G/H]) \cong \mathrm{H}_j(H) \qquad \mathcal{H}(\mathbb{F}_p[G/K]) \cong \mathrm{H}_j(K) \,.$$

Under these identifications, $f_* : \mathrm{H}_j(H) \to \mathrm{H}_j(K)$ is the map induced by the inclusion $H \subset K$ and $\tau_*(f) : \mathrm{H}_j(K; M) \to \mathrm{H}_j(H; M)$ is the associated group homology transfer map [38, 6.3.9] [2, p. 67].

The transfer has the following basic properties, which are easy to verify by calculations with pretransfers. Recall that $\mathcal{H}$ is assumed to commute with direct sums.

10.10. Lemma. *Suppose that $f_1 : S_1 \to T_1$ and $f_2 : S_2 \to T_2$ are maps of G-sets. If f_1 and f_2 are finite-to-one, then so is $f_1 \amalg f_2 : S_1 \amalg S_2 \to T_1 \amalg T_2$, and $\tau_*(f_1 \amalg f_2) = \tau_*(f_1) \oplus \tau_*(f_2)$.*

10.11. Lemma. *Suppose that $f_1 : S_1 \to T$ and $f_2 : S_2 \to T$ are maps of G-sets. If f_1 and f_2 are finite-to-one, the so is $f_1 + f_2 : S_1 \amalg S_2 \to T$, and $\tau_*(f_1 + f_2) = (\tau_*(f_1), \tau_*(f_2))$*

10.12. *Remark.* It follows from 10.9, 10.10, and 10.11 that if $f : S \to T$ is a map of G-sets which is finite-to-one, then $\tau_*(f)$ can be computed in terms of a sum of transfers associated to the projections $G/G_x \subset G/G_{f(x)}$, $x \in S$. For convenience, we will sometimes call such a transfer *the transfer associated to the inclusion $G_x \to G_{f(x)}$.*

10.13. Lemma. *Suppose that $f : S \to T$ and $g : T \to R$ are maps of G-sets. If f and g are finite-to-one then so is $g \cdot f$, and $\tau_*(g \cdot f) = \tau_*(f) \cdot \tau_*(g)$.*

10.14. Lemma. *Suppose that*

$$
\begin{array}{ccc}
S' & \xrightarrow{\ s\ } & S \\
{\scriptstyle f'}\downarrow & & \downarrow{\scriptstyle f} \\
T' & \xrightarrow{\ t\ } & T
\end{array}
$$

is a pullback square of G-sets (i.e. a commutative diagram which induces an isomorphism from S' to the pullback of S and T' over T). Then if f is finite-to-one, so is f', and $\tau_*(f) \cdot t_* = s_* \cdot \tau_*(f')$.

10.15. Definition. A map $f : S \to T$ of G-sets is said to be *an even covering mod p* if it is finite-to-one and the cardinality mod p of $f^{-1}(x)$ does not depend on the choice of $x \in T$. The common value mod p of these inverse image cardinalities is called the *degree* of f and denoted $\deg(f)$.

10.16. Lemma. *Suppose that $f : S \to T$ is a map of G-sets which is an even covering mod p. Then the composite $f_* \cdot \tau_*(f)$ is the endomorphism of $\mathcal{H}(\mathbb{F}_p[T])$ given by multiplication by $\deg(f)$.*

10.17. *Example.* Suppose that H is a subgroup of G and that S is a G-set. The action map $a : G \times_H S \to S$ is finite-to-one and has degree given by the index of H in G. Moreover, if $S' \to S$ is a map of G-sets, the diagram

$$
\begin{array}{ccc}
G \times_H S' & \longrightarrow & G \times_H S \\
{\scriptstyle a}\downarrow & & \downarrow{\scriptstyle a} \\
S' & \longrightarrow & S
\end{array}
$$

is a pullback square. It follows that the maps $\tau_*(a)$ give a natural map

$$\mathcal{H}(\mathbb{F}_p[S]) \xrightarrow{\tau_*(a)} \mathcal{H}(\mathbb{F}_p[G \times_H S])$$

on the category of G-sets. Moreover, the composite $a_* \cdot \tau_*(a)$ is the endomorphism of $\mathcal{H}(\mathbb{F}_p[S])$ given by multiplication by the index of H in G.

> The notation "$G \times_H S$" above stands for the quotient of $G \times S$ by the equivalence relation "$\sim$" given by $(g, hs) \sim (gh, s)$ for $g \in G$, $s \in S$, and $h \in H$. The action of G on this quotient is induced by the action of G on $G \times S$ obtained by setting $g' \cdot (g, s) = (g'g, s)$. There is an isomorphism $\mathbb{F}_p[G \times_H S] \approx \mathbb{F}_p[G] \otimes_{\mathbb{F}_p[H]} \mathbb{F}_p[S]$ of $\mathbb{F}_p[H]$-modules. Show that there is also an isomorphism of G-sets $G \times_H S \approx (G/H) \times S$, where G acts diagonally on the product $(G/H) \times S$.

11. Acyclicity for G-spaces

In this section we translate the question of whether a homology decomposition of BG is sharp into a question about the Bredon homology of the associated G-space (§9). We then study this second question. The symbol $\mathcal{H}$ denotes a coefficient functor (10.1) for G. A lot of the material in this section first appeared (in a slightly different form) in a paper of Webb [37].

11.1. Definition. A G-space X is said to be *acyclic for* $\mathcal{H}$ if the map $X \to *$ induces an isomorphism $H_*^G(X; \mathcal{H}) \to H_*^G(*; \mathcal{H})$.

11.2. Remark. Note that $H_i^G(*; \mathcal{H})$ vanishes for $i > 0$ and $H_0^G(*; \mathcal{H}) = \mathcal{H}(\mathbb{F}_p)$. Let $\mathcal{C}$ be a collection of subgroups of G. By §9 and 10.4, the centralizer, subgroup and normalizer decompositions associated to $\mathcal{C}$ are sharp (8.1) if and only if the G-spaces $X_\mathcal{C}^\alpha$, $X_\mathcal{C}^\beta$, and $X_\mathcal{C}^\delta$ (respectively) are acyclic for the functors $\mathcal{H}_j^G$, $j \geq 0$. In fact, more is true: if any *one* of these three G-spaces is acyclic for all of the functors $\mathcal{H}_j^G$ ($j \geq 0$), then $\mathcal{C}$ is ample, all three spaces correspond to homology decompositions for BG, and the particular decomposition corresponding to the acyclic G-space is sharp.

Note that there are functors

$$\mathbf{X}_\mathcal{C}^\alpha \to \mathbf{X}_\mathcal{C}^\delta \leftarrow \mathbf{X}_\mathcal{C}^\beta$$

which induce G-maps

(11.3) $$X_\mathcal{C}^\alpha \to X_\mathcal{C}^\delta \leftarrow X_\mathcal{C}^\beta .$$

According to the arguments in the proofs of 7.12 and 7.14, these maps are weak equivalences. This reflects the fact that when it comes to the centralizer, subgroup, and normalizer diagrams, either all three give homology decomposition of BG or none of them do. However, the two maps of (11.3) are *not* usually weak G-equivalences. This reflects (10.7) the fact that the three decompositions can have different sharpness properties.

We have two ways to show that a G-space X is acyclic for $\mathcal{H}$. In fact, the second method is a refinement of the first one.

The direct transfer method. This uses the fact that if K is a subgroup of G of index prime to p then the transfer exhibits $\mathrm{H}^G_*(X; \mathcal{H})$ as a retract of $\mathrm{H}^K_*(X; \mathcal{H}|_K)$.

11.4. Theorem. *Suppose that X is a G-space, $\mathcal{H}$ is a coefficient functor for G, and K is a subgroup of G of index prime to p. If X is acyclic as a K-space for $\mathcal{H}|_K$, then X is acyclic as a G-space for $\mathcal{H}$.*

Proof. The transfers (10.8) associated to the maps $q : G \times_K X_n \to X_n$ provide a map $t : C^G_*(X; \mathcal{H}) \to C^K_*(X; \mathcal{H}|_K)$ (9.9). By 10.17 this map commutes with differentials, and the index assumption implies that the composite $C^G_*(X; \mathcal{H}) \xrightarrow{t} C^K_*(X; \mathcal{H}|_K) \xrightarrow{q} C^G_*(X; \mathcal{H})$ is an isomorphism. By naturality, then, the map $\mathrm{H}^G_*(X; \mathcal{H}) \to \mathrm{H}^G_*(*; \mathcal{H})$ is a retract of $\mathrm{H}^K_*(X; \mathcal{H}|_K) \to \mathrm{H}^K_*(*; \mathcal{H}|_K)$, and the theorem follows from the fact that a retract of an isomorphism is an isomorphism. $\square$

11.5. The method of discarded orbits. This is a more sophisticated version of the direct transfer method which exploits the fact that K-orbits can be discarded if they do not contribute to the transfer.

11.6. Theorem. *Let X be a G-space, K a subgroup of G of index prime to p, and Y a subspace of X which is closed under the action of K. Assume that Y is acyclic for $\mathcal{H}|_K$, and that for each simplex $x \in X \setminus Y$ the transfer map $\mathcal{H}(\mathbb{F}_p[G/G_x]) \to \mathcal{H}(\mathbb{F}_p[G/K_x])$ is zero (cf. 10.12). Then X is acyclic for $\mathcal{H}$.*

Proof. The transfers associated to the maps $q : G \times_K X_n \to X_n$ provide a map $t : C^G_*(X; \mathcal{H}) \to C^K_*(X; \mathcal{H}|_K)$ (§10). By 10.17 this map commutes with differentials, and the index assumption implies that the composite $C^G_*(X; \mathcal{H}) \xrightarrow{t} C^K_*(X; \mathcal{H}|_K) \xrightarrow{q} C^G_*(X; \mathcal{H})$ is an isomorphism. The transfer hypothesis shows that the image of t is actually in the subcomplex $C^K_*(Y; \mathcal{H}|_K)$ of $C^K_*(X; \mathcal{H}|_K)$. By naturality, the homology map $\mathrm{H}^G_*(X; \mathcal{H}) \to \mathrm{H}^G_*(*; \mathcal{H})$ is a retract of $\mathrm{H}^K_*(Y; \mathcal{H}|_K) \to \mathrm{H}^K_*(*; \mathcal{H}|_K)$, and the theorem follows from the fact that a retract of an isomorphism is an isomorphism. $\square$

11.7. *Example.* Let X be a G-space, K a subgroup of G of index prime to p, and Y a subspace of X which is closed under the action of K. Suppose that Y is acyclic for the functors $\mathcal{H}^G_j$, $j \geq 0$. Assume finally that for each $x \in X \setminus Y$ the transfer map $\mathrm{H}_*(G_x) \to \mathrm{H}_*(K_x)$ is zero. In light of 10.2, Theorem 11.6 implies that X is acyclic for the functors $\mathcal{H}^G_j$, $j \geq 0$.

A special case of the above is due to Webb [37] [1, V.3]; the proof here is essentially the same as his.

11.8. Corollary. *Let X be a G-space and P a Sylow p-subgroup of G. Suppose that for every nonidentity subgroup Q of P the fixed point space X^Q is contractible, and that for any $x \in X$ there is a (nonidentity) element of order p in G_x. Then X is acyclic for the functors $\mathcal{H}_j^G$, $j \geq 0$.*

Proof. Let Y be the P-subspace of X consisting of simplices which are fixed by a nonidentity element of P. By 10.6, the map $Y \to *$ is a weak P-equivalence, and so by 10.7 the space Y is acyclic for $H_j(P;-)$. Moreover, for any $j \geq 0$ and $x \in X \setminus Y$ the transfer map

$$H_j(G_x) \to H_j(P_x) = H_j(\{e\})$$

is trivial. This is true for $j > 0$ because the target group is zero, and true for $j = 0$ because by assumption the norm map $\sum_{g \in G_x} g : \mathbb{F}_p \to \mathbb{F}_p$ is trivial. The result follows from 11.6 (cf. 11.7). $\qquad\square$

12. Non-identity p-subgroups

Let $\mathcal{C}$ be the collection of all non-identity p-subgroups of G. Assume that $\mathcal{C}$ is nonempty, i.e., that the order of G is divisible by p. In this section we will show that $\mathcal{C}$ is ample and that the three homology decompositions derived from $\mathcal{C}$ are sharp. In all three cases we use the method of Webb (11.8) to show that the spaces $X_{\mathcal{C}}^\delta$, $X_{\mathcal{C}}^\beta$, and $X_{\mathcal{C}}^\alpha$ are acyclic for the coefficient functors $\mathcal{H}_j^G$ ($j \geq 0$) (11.2).

We first recall a property of finite p-groups.

12.1. Lemma. *Any nontrivial finite p-group has a nontrivial center.*

Proof. Let P be a nontrivial finite p-group, and consider the action of P on itself by conjugation. The fixed point set of the action is the center C of P. Since each nontrivial orbit has order a power of p, card(C) is congruent mod p to card(P), so in particular card(C) is divisible by p. Since C certainly contains the identity element, it follows that C also contains nonidentity elements of P. $\qquad\square$

Let P denote a Sylow p-subgroup of G.

12.2. *The normalizer decomposition.* See Webb [37, 2.2.2]. We will show that $X_{\mathcal{C}}^\delta$ is acyclic for the coefficient functors $\mathcal{H}_j^G$. The first step is to show that for each nonidentity subgroup Q of P the space $(X_{\mathcal{C}}^\delta)^Q$ is contractible. By 9.13 and 5.11 $(X_{\mathcal{C}}^\delta)^Q$ is the nerve of the full subcategory **D** of $\mathbf{X}_{\mathcal{C}}^\delta$ generated by the objects H of $\mathbf{X}_{\mathcal{C}}^\delta$ (equivalently, elements $H \in \mathcal{C}$) such that $Q \subset N_G(H)$. The inclusions

$$H \subset H \cdot Q \supset Q$$

provide a zigzag of natural transformations between the identity functor of $\mathbf{D}$ and the constant functor with value Q. The existence of this zigzag implies that $N(\mathbf{D})$ is contractible (5.6).

A typical n-simplex $Q_0 \subset \cdots \subset Q_n$ ($Q_i \in \mathcal{C}$) of $X_{\mathcal{C}}^{\delta}$ has isotropy subgroup $\cap_i N_G(Q_i)$. It is clear that there is an element of order p in this isotropy subgroup; any element of Q_0 will do. Now use 11.8.

12.3. *Remark.* The above result implies that the collection $\mathcal{C}$ of nontrivial p-subgroups of G is ample (see 11.2).

12.4. *The centralizer decomposition.* We again show that $X_{\mathcal{C}}^{\alpha}$ is acyclic for the coefficient functors $\mathcal{H}_j^G$. We first prove that for any nonidentity subgroup Q of P the space $(X_{\mathcal{C}}^{\alpha})^Q$ is contractible. By 9.11 and 5.11, $(X_{\mathcal{C}}^{\alpha})^Q$ is the nerve of the full subcategory $\mathbf{D}$ of $X_{\mathcal{C}}^{\alpha}$ generated by the objects (H, i) with the property that $Q \subset C_G(i(H))$. Let Z be the center of Q and $j : Z \to G$ the inclusion. For an object (H, i) of $\mathbf{D}$, let H' denote the image of the product map $H \times Z \to G$ and $i' : H' \to G$ the inclusion. The maps

$$(H, i) \to (H', i') \leftarrow (Z, j)$$

give a zigzag of natural transformations between the identity functor of $\mathbf{D}$ and the constant functor with value (Z, j). As above, then, $N(\mathbf{D})$ is contractible.

The isotropy subgroup of a typical n-simplex

(12.5) $H_0 \to H_1 \to \cdots H_n \to G$

of $X_{\mathcal{C}}^{\alpha}$ has the form $C_G(Q)$ for some $Q \in \mathcal{C}$. It follows from 12.1 that such an isotropy subgroup contains a nonidentity element of order p. Now use 11.8.

12.6. *The subgroup decomposition.* Again, we show that $X_{\mathcal{C}}^{\beta}$ is acyclic for the coefficient functors $\mathcal{H}_j^G$. As above the first problem is to show that for any non-identity subgroup Q of P the space $(X_{\mathcal{C}}^{\beta})^Q$ is contractible. By 9.12 and inspection $(X_{\mathcal{C}}^{\beta})^Q$ is the nerve of the full subcategory $\mathbf{D}$ of $X_{\mathcal{C}}^{\beta}$ generated by the pairs $(x, G/H)$ with $Q \subset G_x$. The category $\mathbf{D}$ has $(eQ, G/Q)$ as an initial element; in other words, for any object $(x, G/H)$ of $\mathbf{D}$ there is a unique map $(eQ, G/Q) \to (x, G/H)$. This implies (5.4) that $N(\mathbf{D})$ is contractible.

A typical simplex of $X_{\mathcal{C}}^{\beta}$ has as its isotropy subgroup a group of the form Q for some $Q \in \mathcal{C}$. Any such isotropy subgroup contains an element of order p. Now, again, use 11.8.

13. Elementary abelian p-subgroups

In this section we prove several sharpness statements about collections of elementary abelian p-subgroups of G.

Non-identity elementary abelian p-subgroups. Let $\mathcal{C}$ be the collection of all non-identity elementary abelian p-subgroups of G. We will show that $\mathcal{C}$ is both centralizer-sharp and normalizer-sharp. The arguments mimic the ones in §12. Let P be a Sylow p-subgroup of G. In each case we use the method of Webb (11.8) to show that the spaces $X_{\mathcal{C}}^{\delta}$ and $X_{\mathcal{C}}^{\alpha}$ are acyclic for the coefficient functors $\mathcal{H}_j^G$.

Again, we begin with an elementary property of finite p-groups. It is proved by a counting argument like the one in the proof of 12.1

13.1. Lemma. *Let P and Q be finite p-groups with $Q \neq \{e\}$, and suppose that P acts on Q via group automorphisms. Then there exists a nonidentity element x in the center of Q such that x is fixed by the action of P.*

13.2. The normalizer decomposition. Again, see [37, 2.2.2]. We follow 12.2. The first step is to show that for any non-identity subgroup Q of P the space $(X_{\mathcal{C}}^{\delta})^Q$ is contractible. By 9.13 and 5.11, $(X_{\mathcal{C}}^{\delta})^Q$ is the nerve of the full subcategory $\mathbf{D}$ of $\mathbf{X}_{\mathcal{C}}^{\delta}$ determined by the objects H of $\mathbf{X}_{\mathcal{C}}^{\delta}$ (equivalently, elements $H \in \mathcal{C}$) such that $Q \subset N_G(H)$. Let Z be the group of elements of exponent p in the center of Q, and given an object H of $\mathbf{D}$, let H' be the group of elements of exponent p in the center of QH. The inclusions

$$H \supset H \cap H' \subset H'Z \supset Z$$

give a zigzag of natural transformations between the identity functor of $\mathbf{D}$ and the constant functor with value Z. Lemma 13.1 implies that $H \cap H'$ is not the identity subgroup of G. By 5.6, $N(\mathbf{D})$ is contractible.

As in 12.2, the isotropy subgroup of any simplex of $X_{\mathcal{C}}^{\delta}$ contains an element of order p.

13.3. The centralizer decomposition. See [19], where the authors prove a similar theorem for compact Lie groups by a somewhat different argument. We follow the argument of 12.4. The first step is to show that if Q is a non-identity subgroup of P, then $(X_{\mathcal{C}}^{\alpha})^Q$ is contractible. By 9.11 and 5.11, $(X_{\mathcal{C}}^{\alpha})^Q$ is the nerve of the full subcategory $\mathbf{D}$ of $\mathbf{X}_{\mathcal{C}}^{\alpha}$ generated by objects (H, i) such that $Q \subset C_G(i(H))$. Let Z denote group of elements of exponent p in the center of Q and $j : Z \to G$ the inclusion. For an object (H, i) of $\mathbf{D}$, let H' denote the image of the product map $H \times Z \to G$ and $i' : H' \to G$ the inclusion. The maps

(13.4) $$(H, i) \to (H', i') \leftarrow (Z, j)$$

provide a zigzag of natural transformations between the identity functor of $\mathbf{D}$ and the constant functor with value (Z, j). As above, this implies that $N(\mathbf{D})$ is contractible.

As in 12.4, the isotropy subgroup of any simplex of $X_{\mathcal{C}}^{\alpha}$ contains an element of order p.

Smaller collections. If $\mathcal{C}$ is a collection of subgroups of G and K is a subgroup of G, let $\mathcal{C} \cap 2^K$ denote the set of all elements of $\mathcal{C}$ which are subgroups of K. Clearly $\mathcal{C} \cap 2^K$ is a collection of subgroups of K. We are aiming at the following theorem, as an illustration of how it is sometimes possible to work with smaller collections than the collection of all nontrivial elementary abelian p-subgroups.

13.5. Theorem. *Let K be a subgroup of G of index prime to p, and $\mathcal{C}$ a collection of elementary abelian p-subgroups of G. If $\mathcal{C} \cap 2^K$ is centralizer-sharp (as a K-collection) then $\mathcal{C}$ is centralizer-sharp (as a G-collection).*

13.6. *Example.* Theorem 13.5 can be used as a substitute for the argument of 13.3 in showing that the collection $\mathcal{C}$ of all non-identity elementary abelian p-subgroups of G is centralizer-sharp (for the trivial module $\mathbb{F}_p$). To see this, let P be a Sylow p-subgroup of G. It is enough to prove that the collection $\mathcal{C}' = \mathcal{C} \cap 2^P$ of all non-identity elementary abelian p-subgroups of P is centralizer-sharp as a P-collection. We can derive this from 10.7 by showing that $X_{\mathcal{C}'}^\alpha$ is P-equivariantly equivalent to a point. Let $j : Z \to P$ be the inclusion of the group of elements of exponent p in the center of P. If (H, i) is an object of $X_{\mathcal{C}'}^\alpha$, let H' denote the image of the product map $H \times Z \to P$ and $i' : H' \to P$ the inclusion. The maps

$$(H, i) \to (H', i') \leftarrow (Z, j)$$

provide a zigzag of natural transformations between the identity functor of $X_{\mathcal{C}'}^\alpha$ and the constant functor with value (Z, j). This zigzag respects the action of P on $X_{\mathcal{C}'}^\alpha$, and so, according to the arguments of §5, gives an equivariant contraction of $X_{\mathcal{C}'}^\alpha$.

13.7. *Example.* [3] It is possible to do better than the above. Let P be a Sylow p-subgroup of G, and let Z be any non-identity central elementary abelian p-subgroup of P. Let $\mathcal{C}$ be the smallest collection of elementary abelian p-subgroups of G which contains Z and has the property that if $V \in \mathcal{C}$ and V commutes with Z then $\langle Z, V \rangle \in \mathcal{C}$. An argument virtually identical to the one in 13.6 shows that if $\mathcal{C}' = \mathcal{C} \cap 2^P$, then $X_{\mathcal{C}'}^\alpha$ is P-equivariantly contractible. It follows from 13.5 and 10.7 that $\mathcal{C}$ is centralizer-sharp.

The proof of 13.5 depends on the following observation.

13.8. Lemma. *Suppose that K is a subgroup of G and that V is an elementary abelian subgroup of G not entirely contained in K. Then the transfer map*

$$H_*(C_G(V); \mathbb{F}_p) \to H_*(C_G(V) \cap K; \mathbb{F}_p)$$

associated to $C_G(V) \cap K \to C_G(V)$ (see 10.12) is zero.

Proof. Let $C_1 = C_G(V) \cap K$ and $C_2 = C_G(V)$. Choose $v \in V$ with $v \notin K$, and let $C_1' \cong C_1 \times \langle v \rangle$ be the subgroup of C_2 generated by C_1 and v. The inclusion $C_1 \to C_2$ factors as the composite of $f' : C_1' \to C_2$ with $f : C_1 \to C_1'$, so the transfer in question factors (10.13) as a parallel composite $\tau_*(f)\tau_*(f')$. However, the map $\tau_*(f)$ is zero; this follows from the fact that the map

$$f_* : H_*(C_1; \mathbb{F}_p) \to H_*(C_1'; \mathbb{F}_p)$$

is a monomorphism (f has a left inverse) and the fact that the composite $\tau_*(f) \cdot f_*$ is multiplication by p (10.16). $\qquad\square$

Proof of 13.5. Let X be the G-space $X_{\mathcal{C}}^\alpha = N(\mathbf{X}_{\mathcal{C}}^\alpha)$; we have to show that X is acyclic for the functors $H_i(G; -)$. The strategy is to use the method of discarded orbits (11.5). Let $\mathbf{Y}$ be the full subcategory of $\mathbf{X}_{\mathcal{C}}^\alpha$ determined by the objects (H, i) such that $i(H)$ is a subgroup of K, and let $Y = N(\mathbf{Y})$, so that Y is a subspace of $X_{\mathcal{C}}^\alpha$. The action of G on $X_{\mathcal{C}}^\alpha$ restricts to an action of K on Y, and it is clear that Y is equivalent as a K-space to $X_{\mathcal{C}'}^\alpha$, where $\mathcal{C}' = \mathcal{C} \cap 2^K$. In particular, Y is by hypothesis acyclic for the functors $H_i(K; -)$, $i \geq 0$. Let x be a simplex of $X \setminus Y$ as in 12.5, and let V be the image of H_n in G. Since V is not contained in K, Lemma 13.8 guarantees that the homology transfer map associated to the inclusion

$$K_x = C_G(V) \cap K \to G_x = C_G(V)$$

is zero. The desired result follows from 11.7.

14. Appendix

In this section, we sketch the proof of 10.7.

If (X, A) is a pair of G-spaces (i.e., A is a subspace of X), let $C_*^G(X, A; \mathcal{H})$ denote the quotient complex $C_*^G(X; \mathcal{H})/C_*^G(A; \mathcal{H})$ and $H_*^G(X, A; \mathcal{H})$ its homology. It is clear that there is a long exact sequence relating $H_*^G(A; \mathcal{H})$, $H_*^G(X; \mathcal{H})$, and $H_*^G(X, A; \mathcal{H})$.

Let K be a normal subgroup of G. A pair (X, A) is said to be *relatively free mod K* if K acts trivially on the simplices of X not in A and G/K acts freely on these simplices. Let $R = \mathbb{F}_p[G/K]$. If (X, A) is relatively free mod K, then the relative simplicial chain complex $C_*(X, A; \mathbb{F}_p)$ is a chain complex of free R-modules, and there is an evident isomorphism $C_*^G(X, A; \mathcal{H}) \cong \mathcal{H}(R) \otimes_R C_*(X, A)$. The next lemma follows from basic homological algebra.

14.1. Lemma. *Suppose that $f : (X, A) \to (Y, B)$ is a map between pairs of G-spaces which are relatively free mod K, and that f induces an isomorphism $H_*(X, A; \mathbb{F}_p) \cong H_*(Y, B; \mathbb{F}_p)$. Then f induces an isomorphism $H_*^G(X, A; \mathcal{H}) \cong H_*^G(Y, B; \mathcal{H})$.*

Prove this. It follows from the fact that if R is a ring and $f : C \to C'$ is a map of nonnegatively graded chain complexes over R such that

- both C and C' are chain complexes of projective modules, and
- f induces an isomorphism on homology groups,

then f is a chain homotopy equivalence.

We can now prove 10.7. Pick representatives $\{K_i\}_{i=0}^m$ for the conjugacy classes of subgroups of G and label the representatives in such a way that if K_i is conjugate to a subgroup of K_j then $i \geq j$. If Z is a G-space, write $Z^{(n)}$ for the subspace of Z consisting of all $z \in Z$ such that G_z is conjugate to one of the groups K_i for $i \leq n$. Let the height of Z be the least integer n such that $Z = Z^{(n)}$. The proof will be by induction on the heights of the spaces X and Y involved. The result is easy to check if G acts trivially on X and Y, i.e., if the heights of these spaces are ≤ 0.

Assume by induction that the statement of 10.7 is true if the G-spaces involved have height $\leq n-1$. Suppose that X and Y are G-spaces of height $\leq n$ and that $f : X \to Y$ is a map which induces weak equivalences $X^H \to Y^H$ for all subgroups H of G. Let $A = X^{(n-1)}$, $B = Y^{(n-1)}$, $K = K_n$ and $N = N_G(K)$. We must prove that the map $H_*^G(X; \mathcal{H}) \to H_*^G(Y; \mathcal{H})$ is an isomorphism.

It is easy to check that there is a map of pushout squares

$$
\begin{array}{ccc}
G \times_N A^K & \longrightarrow & A \\
\downarrow & & \downarrow \\
G \times_N X^K & \longrightarrow & X
\end{array}
\qquad \longrightarrow \qquad
\begin{array}{ccc}
G \times_N B^K & \longrightarrow & B \\
\downarrow & & \downarrow \\
G \times_N Y^K & \longrightarrow & Y
\end{array}
$$

and that in these squares the vertical arrows are monic. By 10.6 the map $A \to B$ is a weak G-equivalence, so by induction and a long exact sequence argument it is enough to show that the map $H_*^G(X, A; \mathcal{H}) \to H_*^G(Y, B; \mathcal{H})$ is an isomorphism. Given the above diagram of squares, this is equivalent to showing that the map

$$ H_*^N(X^K, A^K; \mathcal{H}|_N) \to H_*^N(Y^K, B^K; \mathcal{H}|_N) $$

is an isomorphism. Since the maps $A^K \to B^K$ and $X^K \to Y^K$ are weak equivalences of spaces, the map $H_*(X^K, A^K; \mathbb{F}_p) \to H_*(Y^K, B^K; \mathbb{F}_p)$ is an isomorphism. The desired result now follows from 14.1, since the N-space pairs (X^K, A^K) and (Y^K, B^K) are relatively free mod K. This last statement follows from the fact that all of the simplices which are added in going from A to X or from B to Y have isotropy group conjugate to K.

Can you use similar ideas to give a proof of 10.6?

References

[1] A. Adem and R. J. Milgram, *Cohomology of finite groups*, Grundlehren der Mathematischen Wissenschaften [Fundamental Principles of Mathematical Sciences], vol. 309, Springer-Verlag, Berlin, 1994.

[2] D. J. Benson, *Representations and cohomology. II*, Cambridge Studies in Advanced Mathematics, vol. 31, Cambridge University Press, Cambridge, 1991, Cohomology of groups and modules.

[3] ———, *Conway's group* co_3 *and the Dickson invariants*, Manuscripta Math. **85** (1994), no. 2, 177–193.

[4] A. K. Bousfield, *Localization and periodicity in unstable homotopy theory*, J. Amer. Math. Soc. **7** (1994), no. 4, 831–873.

[5] ———, *Unstable localization and periodicity*, Algebraic topology: new trends in localization and periodicity (Sant Feliu de Guíxols, 1994), Progr. Math., vol. 136, Birkhäuser, Basel, 1996, pp. 33–50.

[6] A. K. Bousfield and D. M. Kan, *Homotopy limits, completions and localizations*, Springer-Verlag, Berlin, 1972, Lecture Notes in Mathematics, Vol. 304.

[7] K. S. Brown, *Cohomology of groups*, Graduate Texts in Mathematics, vol. 87, Springer-Verlag, New York, 1994, Corrected reprint of the 1982 original.

[8] H. Cartan and S. Eilenberg, *Homological algebra*, Princeton University Press, Princeton, N. J., 1956.

[9] A. Dold and R. Thom, *Quasifaserungen und unendliche symmetrische Produkte*, Ann. of Math. (2) **67** (1958), 239–281.

[10] E. Dror-Farjoun, *Cellular spaces, null spaces and homotopy localization*, Lecture Notes in Mathematics, vol. 1622, Springer-Verlag, Berlin, 1996.

[11] W. G. Dwyer, *Homology decompositions for classifying spaces of finite groups*, Topology **36** (1997), no. 4, 783–804.

[12] ———, *Sharp homology decompositions for classifying spaces of finite groups*, Group representations: cohomology, group actions and topology (Seattle, WA, 1996), Amer. Math. Soc., Providence, RI, 1998, pp. 197–220.

[13] W. G. Dwyer and J. Spaliński, *Homotopy theories and model categories*, Handbook of algebraic topology, North-Holland, Amsterdam, 1995, pp. 73–126.

[14] W. G. Dwyer and C. W. Wilkerson, *A cohomology decomposition theorem*, Topology **31** (1992), no. 2, 433–443.

[15] S. Eilenberg and N. Steenrod, *Foundations of algebraic topology*, Princeton University Press, Princeton, New Jersey, 1952.

[16] P. Gabriel and M. Zisman, *Calculus of fractions and homotopy theory*, Springer-Verlag New York, Inc., New York, 1967, Ergebnisse der Mathematik und ihrer Grenzgebiete, Band 35.

[17] P. G. Goerss and J. F. Jardine, *Simplicial homotopy theory*, Birkhäuser Verlag, Basel, 1999.

[18] J. Grodal, *Higher limits via subgroup complexes*, preprint (MIT) 1999.

[19] S. Jackowski and J. McClure, *Homotopy decomposition of classifying spaces via elementary abelian subgroups*, Topology **31** (1992), no. 1, 113–132.

[20] S. Jackowski, J. McClure, and B. Oliver, *Homotopy classification of self-maps of BG via G-actions. I*, Ann. of Math. (2) **135** (1992), no. 1, 183–226.

[21] D. M. Kan and W. P. Thurston, *Every connected space has the homology of a $K(\pi, 1)$*, Topology **15** (1976), no. 3, 253–258.

[22] S. Mac Lane, *Categories for the working mathematician*, Springer-Verlag, New York, 1971, Graduate Texts in Mathematics, Vol. 5.

[23] ――――, *Homology*, Classics in Mathematics, Springer-Verlag, Berlin, 1995, Reprint of the 1975 edition.

[24] J. P. May, *Simplicial objects in algebraic topology*, Chicago Lectures in Mathematics, University of Chicago Press, Chicago, IL, 1992, Reprint of the 1967 original.

[25] D. McDuff, *On the classifying spaces of discrete monoids*, Topology **18** (1979), no. 4, 313–320.

[26] J. R. Munkres, *Elements of algebraic topology*, Addison-Wesley Publishing Company, Menlo Park, Calif., 1984.

[27] D. G. Quillen, *Homotopical algebra*, Springer-Verlag, Berlin, 1967, Lecture Notes in Mathematics, No. 43.

[28] ――――, *The geometric realization of a Kan fibration is a Serre fibration*, Proc. Amer. Math. Soc. **19** (1968), 1499–1500.

[29] ――――, *On the (co-) homology of commutative rings*, Applications of Categorical Algebra (Proc. Sympos. Pure Math., Vol. XVII, New York, 1968), Amer. Math. Soc., Providence, R.I., 1970, pp. 65–87.

[30] ――――, *On the cohomology and K-theory of the general linear groups over a finite field*, Ann. of Math. (2) **96** (1972), 552–586.

[31] ――――, *Higher algebraic K-theory. I*, (1973), 85–147. Lecture Notes in Math., Vol. 341.

[32] E. H. Spanier, *Algebraic topology*, Springer-Verlag, New York, 1981, Corrected reprint.

[33] N. E. Steenrod, *A convenient category of topological spaces*, Michigan Math. J. **14** (1967), 133–152.

[34] J. Thévenaz and P. Webb, *The structure of Mackey functors*, Trans. Amer. Math. Soc. **347** (1995), no. 6, 1865–1961.

[35] R. W. Thomason, *Homotopy colimits in the category of small categories*, Math. Proc. Cambridge Philos. Soc. **85** (1979), no. 1, 91–109.

[36] P. J. Webb, *A local method in group cohomology*, Comment. Math. Helv. **62** (1987), no. 1, 135–167.

[37] ———, *A split exact sequence of Mackey functors*, Comment. Math. Helv. **66** (1991), no. 1, 34–69.

[38] C. A. Weibel, *An introduction to homological algebra*, Cambridge Studies in Advanced Mathematics, vol. 38, Cambridge University Press, Cambridge, 1994.

[39] T. Yoshida, *On G-functors. II. Hecke operators and G-functors*, J. Math. Soc. Japan **35** (1983), no. 1, 179–190.

Department of Mathematics,
University of Notre Dame,
Notre Dame,
IN 46556 USA
E-mail address: `dwyer.1@nd.edu`

Cohomology of Groups and Unstable Modules over the Steenrod Algebra*

Hans-Werner Henn

0. Introduction

In these notes we describe how recent developments in the theory of unstable modules over the Steenrod algebra can be used to study the mod-p cohomology ring H^*BG of a group G. These developments rely heavily on properties of Lannes' functor T_V and they can be applied even more generally to study the mod-p cohomology of Borel constructions. The groups for which the techniques work need to satisfy some finiteness conditions; in particular their mod-p cohomology ring needs to be a finitely generated algebra. For example, all compact Lie groups (including finite groups) qualify, but also many classes of discrete groups like arithmetic groups (and more generally S-arithmetic groups), mapping class groups, automorphism groups of free groups .... Another case of interest is the continuous mod-p cohomology of p-adic Lie groups.

In section 1 we start off by recalling some fundamental results in cohomology of groups, in particular the Evens-Venkov result on finite generation of the cohomology ring (Theorem 2) and Quillen's landmark result which describes H^*BG up to F-isomorphism (Theorem 5). In section 2 we introduce unstable modules and unstable algebras over the Steenrod algebra and we give an interpretation of Quillen's F-isomorphism theorem in terms of unstable modules (Theorem 11). The theory of Lannes' functor T_V is sketched in section 3 and his computation of $T_V H^*(BG)$ in terms of the cohomology of centralizers of elementary abelian p-subgroups (Theorem 19) is explained. The interpretation of Quillen's result given in section 2 turns out to be equivalent to the computation of the degree 0 part of the unstable module $T_V H^*BG$.

---

*These notes, which elaborate on parts of a survey article with the title "Unstable modules over the Steenrod algebra and cohomology of groups" [H6], are a revised version of the notes for an advanced course which was held at the CRM in Bellaterra from May 27 to June 2, 1998. I would like to thank the participants for numerous discussions. Several suggestions made in these discussions have been incorporated in this version in some way or another. I would like to explicitly thank Jon Berrick for his critical reading of the first version of these notes.

Lannes' computation of $T_V H^* BG$ suggests a strategy for a proof of Quillen's original theorem which is also presented in this section. This proof uses essentially the same ideas as Quillen's, but in a streamlined form. Given that Quillen's theorem is a consequence of just the degree 0 part of Lannes' computation, one wonders what the rest of the computation of $T_V H^* BG$ means for the cohomology of groups. In a way, the remaining three sections of these notes adress this point.

Quillen's result is a "computation" of $H^* BG$, up to nilpotent modules, in terms of the cohomology of the elementary abelian p-subgroups of G. In the world of unstable modules one can refine the notion of nilpotent modules and introduce n-nilpotent modules (Definition 7 – Proposition 20); a nilpotent module is nothing but a 1-nilpotent module, and every $(n+1)$-nilpotent module is also n-nilpotent. Lannes' computation of $T_V H^* BG$ allows to "compute" $H^* BG$, up to n-nilpotent modules, in terms of the cohomology of the elementary abelian p-subgroups of G and the cohomology of their centralizers in degree less than n (Theorem 22). One obtains in this way a sequence of approximations $L_n H^* BG$ of $H^* BG$ which for large n agree with $H^* BG$. Furthermore, there are explicit bounds which tell us when n is large enough (Theorem 24). All this is discussed in section 4. The proper framework for this discussion is localization in abelian categories which we also review in this section. In section 5 we analyze how much of $H^* BG$ is determined if we know the mod p cohomology of (suitable) collections of centralizers of elementary abelian p-subgroups of G (Theorem 30 and Corollary 31). If G is a compact Lie group and we take the collection of all non-trivial elementary abelian p-subgroups, we obtain a celebrated result of Jackowski and McClure. Finally, in section 6 we discuss a centralizer spectral sequence which is associated to a homotopy colimit decomposition of Borel constructions. We describe a general strategy how this can be used to study the cohomology of suitable discrete groups. This strategy works extremely well in the case of the group $SL(3, \mathbb{Z}[1/2])$ and in the case of certain orthogonal groups over $\mathbb{Z}[1/2]$; in both cases it leads to complete and explicit calculations of the mod-2 cohomology ring. One can expect that there are other interesting examples waiting to be discovered.

These notes are to a large extent expository. I have tried to give at least outlines or ideas of proofs and describe the general philosophy, but for complete details the reader will often have to consult the original sources. Nevertheless I hope that these notes serve as a guide to those which want to learn more about the subject.

1. Cohomology of groups

1.1. Some fundamental structural results for cohomology groups and cohomology rings of finite groups

For a topological group G we denote its classifying space by BG. Our groups will usually be either discrete or compact Lie so that up to homotopy equivalence there is no difference between Milnor's infinite join construction or the construction of BG as the classifying space of a (topological) category with a single object and whose morphisms are given by G. If the group is discrete then the cohomology of BG with local coefficients in a G-module M is also known as the cohomology of the group G with coefficients in M (and is usually denoted $H^*(G; M)$). In these notes the action of G on M will usually be trivial and we will always use the topologists notation $H^*(BG; M)$. We will begin by recalling some basic properties of the cohomology of finite groups.

Our first result is a finiteness result. Its proof relies heavily on elementary properties of the transfer.

Proposition 1. *Let G be a finite group. Then $H^n(BG; \mathbb{Z})$ is a finitely generated group for all n and multiplication with the order of G annihilates $H^n(BG; \mathbb{Z})$ for all $n > 0$.*

Example. Let $\mathbb{Z}/k$ denote the cyclic group of order k. Then $H^{2n}(B\mathbb{Z}/k; \mathbb{Z}) \cong \mathbb{Z}/k$ if $n > 0$ and $H^{2n+1}(B\mathbb{Z}/k; \mathbb{Z}) = 0$.

The proposition focuses interest on studying p-torsion in $H^*(BG; \mathbb{Z})$ for those primes p which divide the order of G and to the closely related cohomology ring $H^*(BG; \mathbb{F}_p)$. In fact, we will almost exclusively discuss these rings. However, the first fundamental result holds for more general coefficients.

Theorem 2 [E1], [V]. *Let G be a finite group and R be a noetherian commutative ring. Then $H^*(BG; R)$ is a finitely generated (graded commutative) R-algebra.*

One way of proving this uses an embedding of G into $U(n)$, the group of unitary $n \times n$ matrices, and knowledge of $H^*(BU(n); \mathbb{Z})$. We recall that $H^*(BU(n); \mathbb{Z}) \cong \mathbb{Z}[c_1, \ldots c_n]$ where the elements c_i are the universal Chern classes and live in degree $2i$. (The strategy to use compact Lie groups to prove theorems for finite groups will be used again in Section 3 below!) Theorem 2 is now an easy consequence of the following result.

Theorem 3. *Let G be a compact Lie group, H a closed subgroup and R a noetherian ring. Then $H^*(BH; R)$ becomes a finitely generated $H^*(BG; R)$-module via the map induced by the inclusion of H into G.*

Outline of the proof of Theorem 3. We choose an embedding of G into $U(n)$ (for example the regular representation if G is finite). It suffices to prove the theorem in case $G = U(n)$. Consider the $H^*(-; R)$-based Serre spectral sequence of the fibration $BH \longrightarrow BU(n)$, with fibre $F = U(n)/H$. This is a spectral sequence of $H^*(BU(n); R) \cong R[c_1, \dots, c_n]$-modules and its E_2-term is finitely generated over this ring because the cohomology of the fibre (which is a compact manifold) is a finitely generated R-module. Because $R[c_1, \dots, c_n]$ is noetherian it follows first that all other pages of the spectral sequence consist of finitely generated $H^*(BU(n); R)$-modules and consequently the same holds for $H^*(BH; R)$. $\square$

Here is another very useful consequence of this result.

Corollary 4. *Let G be a finite group and assume G contains an element of order p^k. Then there exists $n > 0$ and an element $z \in H^n(BG; \mathbb{Z})$ such that the order of each power of z is divisible by p^k.*

Proof. An element of order p^k defines an embedding ι of $\mathbb{Z}/p^k$ into G. Now use that as a ring $H^*(B\mathbb{Z}/p^k; \mathbb{Z})$ is isomorphic to $\mathbb{Z}[y]/(p^k y)$ with y in degree 2. By Theorem 3 $H^*(B\mathbb{Z}/p^k; \mathbb{Z})$ becomes a finitely generated module over $H^*(BG; \mathbb{Z})$ via the map induced by ι and this is only possible if some power y^k of y is in the image of this map. A preimage of this power qualifies as z. $\square$

As a variation of this proof one could calculate the Chern classes of the regular representation of G restricted to the subgroup generated by the element of order p^k. This has the advantage of giving a concrete class z with properties as in Corollary 4. What degree is it in?

The case of the group $G = \mathbb{Z}/p \times \mathbb{Z}/p$ shows that in general there is no element in $H^*(BG; \mathbb{Z})$ whose order is equal to the order of the group. If G is an arbitrary finite group then finding the maximal order of elements of positive degree of the ring $H^*(BG; \mathbb{Z})$ (this order is often called the *exponent* of $H^*(BG; \mathbb{Z})$) is an interesting and difficult open problem.

1.2. Quillen's F-isomorphism theorem

Theorem 2 suggests to use the usual concepts of commutative algebra in order to study $H^*(BG; R)$. Of course, one would like to understand these concepts in terms of the group theory of G. Quillen's fundamental result [Q1] takes a substantial step in this direction if $R = \mathbb{F}_p$; it describes $H^*(BG; \mathbb{F}_p)$ up to nilpotency phenomena and determines the Krull dimension of $H^*(BG; \mathbb{F}_p)$ and its variety (or rather its set of geometric points) in a beautiful and very satisfactory way. Quillen's approach was to understand as much as possible about $H^*(BG; \mathbb{F}_p)$ by relating it via the restriction maps to the cohomology of its elementary abelian p-subgroups, i.e. groups isomorphic to $(\mathbb{Z}/p)^n$ for some natural number n. We recall that for an elementary abelian p-group E we have

$$H^*(BE; \mathbb{F}_p) \cong \begin{cases} S(E^*, 1) & \text{if } p = 2, \\ S(E^*, 2) \otimes \Lambda(E^*, 1) & \text{if } p > 2. \end{cases}$$

Here E^* is the dual of E and $S(E^*, i)$ stands for a graded symmetric algebra with E^* in degree i and similarly $\Lambda(E^*, i)$ for a graded exterior algebra with E^* in degree i.

Now consider the map $H^*(BG; \mathbb{F}_p) \longrightarrow \prod_E H^*(BE; \mathbb{F}_p)$ with E running through all elementary abelian p-subgroups of G. It is clear that elements in the target of this map must satisfy compatibility conditions (arising from inclusions and conjugations in G) to have a chance to be in its image.

More precisely Quillen considered the following category $\mathcal{A}_p(G)$: its objects are the elementary abelian p-subgroups of G and its morphisms are all group homomorphisms which can be induced by conjugation by an element in G; i.e. for E_1 and E_2 in $\mathcal{A}_p(G)$ one defines

$$\text{mor}_{\mathcal{A}_p(G)}(E_1, E_2) = \{\alpha \in \text{Hom}(E_1, E_2) | \exists g \in G$$
$$\text{such that } \alpha(e) = geg^{-1} \forall e \in E_1\} \, .$$

Then $E \mapsto H^*(BE; \mathbb{F}_p)$ becomes a (covariant) functor from the opposite category $\mathcal{A}_p(G)^{op}$ to graded $\mathbb{F}_p$-algebras and the restriction homomorphisms induce a map $q_G : H^*(BG; \mathbb{F}_p) \longrightarrow \lim_{\mathcal{A}_p(G)^{op}} H^*(BE; \mathbb{F}_p)$. Quillen's result reads now as follows.

Theorem 5 [Q1]. *Let p be a prime and G be a finite group. Then the map*

$$q_G : H^*(BG; \mathbb{F}_p) \longrightarrow \lim_{\mathcal{A}_p(G)^{op}} H^*(BE; \mathbb{F}_p)$$

has the following properties:

- *If $x \in \text{Ker} \, q_G$ then x is nilpotent.*
- *For every $y \in \lim_{\mathcal{A}_p(G)^{op}} H^*(BE; \mathbb{F}_p)$ there exists a natural number n with $y^{p^n} \in \text{Im} \, q_G$.*

A map of graded $\mathbb{F}_p$-algebras satisfying the two conditions occuring in the theorem is called an *F-isomorphism*. We will give an outline of a proof of Quillen's Theorem in Section 3.3 below.

Quillen [Q1] showed that Theorem 5 is true for other classes of groups as well, e.g.

a) if G is a discrete group of finite virtual mod p-cohomological dimension. (It will not be important for these notes to know the precise definition of this class of groups. However, we emphasize that this class includes many "classical" groups like all (S-) arithmetic groups, e.g. $SL(n, \mathbb{Z})$, mapping class groups of Riemann surfaces, automorphism groups of free groups ..., i.e. the symmetry groups of some of the most basic objects in mathematics.)

b) if G is a compact Lie group

c) if G is a profinite group and $H^*(BG; \mathbb{F}_p)$ is replaced by the continuous mod-p cohomology $H^*_{cts}(G; \mathbb{F}_p)$ and is assumed to be noetherian (this includes all p-adic analytic groups in the sense of Lazard [Lz]).

Definition 1. A topological group G is called a *Quillen group* at the prime p if the map q_G is an F-isomorphism.

In other words, all compact Lie groups and all discrete groups of finite virtual mod p-cohomological dimension are Quillen groups. There is an obvious modification of the definition in the context of continuous cohomology of profinite groups but we will not pursue this here. For other classes of discrete Quillen groups we refer to [H2] and [H6].

Example. To construct discrete groups that are not Quillen groups takes some effort. For example, the class of Quillen groups cannot be expected to be closed with respect to colimits: the cohomology $H^*(BG; \mathbb{F}_p)$ of a group G, which is the colimit of a sequence of homomorphisms $G_n \to G_{n+1}$, is the limit $\lim_n H^*(BG_n; \mathbb{F}_p)$, and in this way one may get non-nilpotent elements in $H^*(BG; \mathbb{F}_p)$ which restrict trivially on any elementary abelian p-subgroup. An explicit example may be constructed as follows: according to [AC] resp. [IK] there exist finite 2-groups F_n with elements $t_n \in H^1(BF_n; \mathbb{F}_2)$ such that $t_n^n \neq 0$, $t_n^{n+1} = 0$. Let $G_n = \prod_{k \leq n} F_k$ and let G be the colimit with respect to thecanonical inclusions of G_n into G_{n+1}.

Excercise 1: Show that there is a non-nilpotent class in $H^1(BG; \mathbb{F}_2)$.

The class of Quillen groups is not well understood, in fact so far it has hardly been investigated for its own sake. It would be interesting to have a characterization of Quillen groups in more group theoretical terms.

We give some of the immediate applications of Theorem 5. The first one determines the Krull dimension of the ring $H^*(BG; \mathbb{F}_p)$ in terms of the group theory of G, more precisely in terms of $r_p(G)$, the p-rank of G, which is defined as the maximal rank of an elementary abelian p-subgroup of G.

For our purpose it is convenient to define the Krull dimension as follows: if M is a finitely generated module over a connected finitely generated $\mathbb{F}_p$-algebra, then the power series $\chi_M(t) = \sum_n \dim_{\mathbb{F}_p} M^n t^n$ can be shown to be the power series expansion of a rational function in t of the form

$$\frac{p(t)}{\prod_{i=1}^k (1 - t^{n_i})},$$

where $p(t)$ is a polynomial in t with integer coefficients. The Krull dimension $\delta(M)$ of M is then defined to be equal to the order of the pole of the power series $\chi_M(t)$ at $t = 1$. If G is a finite group and p is a prime we set $\chi_{G,p}(t) = \chi_{H^*(BG;\mathbb{F}_p)}(t)$ and $\delta_p(G) = \delta(H^*(BG;\mathbb{F}_p))$.

This means that the Krull dimension measures the growth of the series of numbers $\dim_{\mathbb{F}_p} M^n$. In fact, it is a nice exercise on power series of rational functions to show that the Krull dimension of an M as above is the smallest integer δ such that there exists a constant C with $\dim_{\mathbb{F}_p} M^n \le Cn^{\delta-1}$ for all $n \ge 1$.

Example. If $V \cong (\mathbb{Z}/p)^n$ then $\chi_{V,p}(t) = \frac{1}{(1-t)^n}$, hence $\delta_p(V) = n$.

Theorem 6. *If G is any finite group then $\delta_p(G) = r_p(G)$.*

Sketch of proof. The inequality $r_p(G) \le \delta_p(G)$ is an easy consequence of Theorem 3 applied to the inclusion of an elementary abelian p-subgroup E into G.

For the other inequality let $I = \ker q_G$ and $L = \lim_{A_p(G)^{op}} H^*(BE;\mathbb{F}_p)$ so that we have an embedding $H^*(BG;\mathbb{F}_p)/I \hookrightarrow L$. Then it is easy to see that $\delta(H^*(BG;\mathbb{F}_p)/I) \le \delta(L) \le r_p(G)$. The ideal I is finitely generated and nilpotent hence $I^N = 0$ for some N. Moreover, I^j/I^{j+1} is a finitely generated $H^*(BG;\mathbb{F}_p)/I$-module and from this one deduces $\delta(I^j/I^{j+1}) \le \delta(H^*(BG)/I) \le r_p(G)$ and finally $\delta_p(G) \le r_p(G)$. $\square$

Quillen used Theorem 5 also to describe the variety of $H^*(BG;\mathbb{F}_p)$, more precisely its geometric points with values in an algebraically closed field k, in group theoretic terms and thus to lay the basis for a very interesting subsequent development in modular representation theory, namely the theory of varieties attached to modules. We will not pursue this here and refer instead to [Be] and [E2] for more information on this development.

Theorem 5 can also be used to construct interesting elements in

$$H^*(BG;\mathbb{F}_p).$$

For example, let D be a maximal elementary abelian p-subgroup of G.

If $p = 2$ let

$$c_D := \prod_{0 \ne x \in H^1(BE;\mathbb{F}_p)} x,$$

and if p is odd let

$$c_D := \prod_{0 \ne x \in H^1(BE;\mathbb{F}_p)} \beta x$$

(where β is the Bockstein operator), be the top Dickson invariant in $H^*(BD;\mathbb{F}_p)$. This class is invariant with respect to the natural action of the general linear group $GL(D)$ on $H^*(BD;\mathbb{F}_p)$ and it restricts trivially to

$H^*(BD'; \mathbb{F}_p)$ for each subgroup D' of D: hence the family of elements $(x_E) \in$ $\prod_{E \in \mathcal{A}_p(G)} H^*(BE; \mathbb{F}_p)$, given by $x_E = c_E$ if E is conjugate to D and $x_E = 0$ otherwise, defines an element y_D in $\lim_{\mathcal{A}_p(G)^{op}} H^*(BE; \mathbb{F}_p)$, and by Theorem 5 there is an integer k and an element $z_D \in H^*(BG; \mathbb{F}_p)$ such that $q_G(z_D) = y_D^{p^k}$.

Corollary 7. *Let G be a finite group and assume there are two maximal elementary abelian p-subgroups of G which are not conjugate. Then there are non-nilpotent elements z_1 and z_2 in $H^*(BG; \mathbb{F}_p)$ such that $z_1 z_2 = 0$.*

Proof. Let D_1 and D_2 be two such subgroups and construct elements z_1' and z_2' as above. Then $q_G(z_1' z_2') = 0$, hence $z_1' z_2'$ is nilpotent. If we replace z_i' for $i = 1, 2$ by a sufficiently large power we get the desired elements z_i. □

For a refinement of this statement we refer to Proposition 29 below.

Examples. We finish this section with some examples illustrating Theorem 5.

a) The map q_G is not an isomorphism in general. The first example is given by the cyclic groups $\mathbb{Z}/p^k$, $k > 1$, in which case the map is neither injective nor surjective.

b) However, there are interesting classes of groups G for which q_G is an isomorphism or at least a monomorphism. For example, if $p = 2$, q_G is an isomorphism if $G = D_{2^n}$, the dihedral group of order 2^n, or if $G = \mathfrak{S}_n$, the symmetric groups on n letters, or $G = O_n(\mathbb{F}_q)$, the symmetry group of the bilinear form $b(x, y) = \sum_i x_i y_i$ on $\mathbb{F}_q^n$ if $\mathbb{F}_q$ is a field of odd order. (The general phenomenon underlying these examples is the following: the property that q_G is an isomorphism is preserved by passing from a group G to its wreath product with $\mathbb{Z}/2$, and is inherited by any group whose Sylow subgroup has this property (see [GLZ]).)

c) Let D_8 be the dihedral group of order 8. It is known that $H^*(D_8; \mathbb{F}_2) \cong$ $\mathbb{F}_2[x_1, y_1, w_2]/(x_1 y_1)$ where the indices give the degree of the elements (cf. [AM]). The elements x_1 and y_1 are two classes as in Corollary 7, and this reflects the fact that D_8 has two non-conjugate elementary abelian 2-subgroups of rank 2.

In general Theorem 5 says absolutely nothing about nilpotent elements. In later sections we will see that Lannes' generalization of Quillen's Theorem (cf. Theorem 19 below) provides information on nilpotent elements as well.

2. Quillen's F-isomorphism theorem from the point of view of unstable modules

2.1. Unstable modules over the Steenrod algebra

We will see below that Quillen's theorem can be reinterpreted as a statement on $H^*(BG; \mathbb{F}_p)$ considered as an unstable module over the Steenrod algebra. Before we get to this we review some of the basic facts on Steenrod operations and unstable modules. For details we refer to [SE] and [Sc].

Definition 2. *Let p be a prime. The mod-p Steenrod algebra A_p is the following graded associative $\mathbb{F}_p$-algebra:*

If $p = 2$ it is generated by elements Sq^i, $i = 0, 1, 2, \ldots$ of degree i, with $Sq^0 = 1$ and subject to the following Adem relations:

$$\text{if } 0 < a < 2b: \ Sq^a Sq^b = \sum_{c=0}^{[a/2]} \binom{b-c-1}{a-2c} Sq^{a+b-c} Sq^c .$$

If p is odd it is generated by elements P^i, $i = 0, 1, 2, \ldots$ of degree $2i(p-1)$, with $P^0 = 1$, together with an element β of degree 1 with $\beta^2 = 0$, and subject to the following Adem relations:

$$\text{if } 0 < a < pb: \ P^a P^b = \sum_{c=0}^{[a/p]} (-1)^{a+c} \binom{(p-1)(b-c)-1}{a-pc} P^{a+b-c} P^c ,$$

$$\text{if } 0 < a \le pb: \ P^a \beta P^b = \sum_{c=0}^{[a/p]} (-1)^{a+c} \binom{(p-1)(b-c)}{a-pc} \beta P^{a+b-c} P^c +$$

$$+ \sum_{c=0}^{[(a-1)/p]} (-1)^{a+c-1} \binom{(p-1)(b-c)-1}{a-pc-1} P^{a+b-c} \beta P^c .$$

Fortunately we do not have to explicitly work with the Adem relations. What is important for us are some formal properties of the category $\mathcal{U}$ of unstable modules (see Definition 3 below). Of course, the Adem relations are crucial in establishing these properties in the first place.

The importance of the Steenrod algebra comes from the fact that mod-p cohomology is a functor from the category $\mathcal{T}op$ of topological spaces to the category of graded A_p-modules. In fact, more is true: the cohomology of a space is a graded A_p-module with a product stucture (given by the cup-product) for which the action satisfies additional "instability axioms", axioms which define the categories of unstable modules resp. unstable algebras over the mod-p Steenrod algebra.

Definition 3. *An unstable A_p-module M is a graded A_p-module such that*

- $Sq^i x = 0$ *for all $i > |x|$, if $p = 2$;*
- $\beta^e P^i x = 0$ *for all $2i + e > |x|$, if p is odd and $e \in \{0, 1\}$.*

Here and elsewhere in this paper $|x|$ denotes the degree of the element x.

As usual we will write $\mathcal{U}_p$ for the (abelian) category of unstable modules and A_p-linear maps. Usually p will be fixed and understood from the context and then we will use the abbreviations *unstable module* and $\mathcal{U}$ resp.

Definition 4.   *An unstable A_p-algebra is an unstable A_p-module K together with linear maps $\varphi : K \otimes K \longrightarrow K$, $\eta : \mathbb{F}_p \longrightarrow K$ such that*

- *φ and η determine a graded commutative $\mathbb{F}_p$-algebra structure on K.*

- *φ is A_p-linear. (For this to make sense we recall that A_p is a Hopf algebra, hence the category of graded A_p-modules has an associated internal tensor product. Another more explicit way of stating this is that the product in K satisfies the Cartan formula, i.e. $Sq^n(xy) = \sum_{i+j=n} Sq^i x Sq^j y$ if $p = 2$; $\beta(xy) = \beta(x)y + (-1)^{|x|}x\beta(y)$, $P^n(xy) = \sum_{i+j=n} P^i x P^j y$ if p is odd.)*

-
$$Sq^n x = x^2 \text{ if } p = 2 \text{ and } n = |x|,$$
$$P^n x = x^p \text{ if } p > 2 \text{ and } 2n = |x| \ .$$

As usual we will write $\mathcal{K}_p$ for the category of unstable algebras over A_p; its objects are the unstable algebras and its morphisms the degree preserving maps which are both A_p-linear and maps of graded algebras. As above we will often use the abbreviations *unstable algebra* and $\mathcal{K}$ if p is understood from the context.

As mentioned before, the mod-p cohomology of a space, in particular the cohomology of the classifying space of a group, i.e. the cohomology of a group, is a natural example of an unstable algebra, the product φ being given by the cup product.

Example.   Let $p = 2$ and $V = \mathbb{Z}/p^n$. Then $H^*(BV; \mathbb{F}_2) \cong \mathbb{F}_2[x_1, \cdots x_n]$ where the degree of the elements x_i is 1. The action of the Steenrod algebra is determined by the instability axioms. By the Cartan formula it suffices to determine $Sq^n x_i$ if $i > 0$ and the instability axioms give $Sq^1 x_i = x_i^2$ while $Sq^n x_i = 0$ if $n > 1$. The case of an odd prime is similar and left as an exercise.

So the action of the Steenrod algebra on $H^*(BV; \mathbb{F}_p)$ is explicitly understood. Nevertheless $H^*(BV; \mathbb{F}_p)$ remains a complicated object. To test your understanding here are some exercises.

Excercise 2:

a) Show that the endomorphism ring $End_{\mathcal{U}_2}(\widetilde{H}^*(B\mathbb{Z}/2; \mathbb{F}_2))$ of the reduced homology $\widetilde{H}^*(B\mathbb{Z}/2; \mathbb{F}_2)$ is isomorphic to $\mathbb{F}_2$, in other words every nontrivial endomorphism is equal to the identity. (Hint: It is enough to use the action of Sq^1 and Sq^2.)

b) What can you say about $End_{\mathcal{U}_p}(\widetilde{H}^*(B\mathbb{Z}/p; \mathbb{F}_p))$?

The conceptual importance of the cohomology of an elementary abelian p-group is underlined by the following deep result of Carlsson resp. Miller which in the form given here is due to Lannes and Zarati. After having struggled with Exercise 2 above you will probably appreciate the depth and beauty of this result; in any case this result was the beginning of a flourishing period of research on unstable modules and homotopy theory of classifying spaces and is also crucial for most of the results we will discuss in these notes.

Theorem 8 [C], [M], [LZ]. *Let V be an elementary abelian p-group. Then $H^*(BV; \mathbb{F}_p)$ is an injective object in the category $\mathcal{U}_p$.*

Comments on the proof. The proof is quite indirect. First one considers the case that the rank of V is 1. One starts from some tautologically injective objects, the Brown Gitler modules $J(n)$ which represent the (exact) functor from $\mathcal{U}_p$ to the category $\mathcal{E}_\infty$ of $\mathbb{F}_p$-vector spaces which sends an unstable module M to $(M^n)^*$, the linear dual of the $\mathbb{F}_p$-vector space of homogeneous elements of M of degree n. From the Brown Gitler modules one constructs other injectives, the Carlsson modules $K(i)$ which for $p = 2$ are given as $\lim_i J(2^i)$ and represent the exact functor $\mathcal{U}_2 \to \mathcal{U}_2, M \mapsto (\operatorname{colim}_k M^{2^k i})^*$ where the maps in the system are given by $M^{2^k i} \longrightarrow M^{2^{k+1} i}, x \mapsto Sq^{|x|} x$. The hard step is then to show that $H^*(B\mathbb{Z}/p; \mathbb{F}_p)$ is a direct summand in $K(1)$.

The case of a general V is dealt with by proving that for any injective I the tensor product $I \otimes H^*(B\mathbb{Z}/p; \mathbb{F}_p)$ is again injective. (This step uses the Carlsson modules again.) We refer to the original papers or to [Sc] for the details of the proof. □

Warning: It is not true that the tensor product of any two injectives remains injective.

We remark that the finite groups G for which $H^*(BG; \mathbb{F}_p)$ is injective in $\mathcal{U}_p$ are precisely those whose p-Sylow subgroup is elementary abelian [H1].

Before we can connect Theorem 5 with the theory of unstable modules we need to define nilpotent unstable modules.

Definition 5. An unstable A_p-module N is called *nilpotent* iff for each $x \in N$ there exists a natural number n such that

- $Sq^{2^n|x|} \cdots Sq^{|x|} x = 0$, if $p = 2$;
- $P^{p^n \frac{|x|}{2}} \cdots P^{\frac{|x|}{2}} x = 0$, if $p > 2$ and x has even degree.

The justification for this terminology comes from the last property defining an unstable algebra, i.e. from the fact that in this case the condition reads $x^{p^{n+1}} = 0$.

If p is odd, one has to be a bit careful with this notion. The unstable algebra $H^*(B\mathbb{Z}/p; \mathbb{F}_p)$ has elements which are nilpotent in the classical sense (the radical, i.e. the ideal of nilpotent elements, consists precisely of the elements of odd degree) but it does not have any non-trivial nilpotent unstable submodules.

Excercise 3: Show that if $p = 2$, or if p is odd and $\beta = 0$ on K, then the ideal of nilpotent elements in an unstable algebra K is a nilpotent submodule.

The following important result of Lannes and Schwartz gives a characterization of nilpotent modules in terms of the injective objects $H^*(BV; \mathbb{F}_p)$.

Theorem 9 [LS]. *An unstable module N is nilpotent if and only if*

$$\text{Hom}_{\mathcal{U}_p}(N, H^*(BV; \mathbb{F}_p)) = 0$$

for all elementary abelian p-groups V.

Comments on the proof. The easy part is to show that there are no non-trivial homomorphisms from a nilpotent unstable module to $H^*(BV; \mathbb{F}_p)$. If $p = 2$ this is clear because $H^*(BV; \mathbb{F}_2)$ has no non-trivial nilpotent elements and the map $x \mapsto Sq^{|x|}x$ is the squaring map and hence injective on $H^*(BV; \mathbb{F}_2)$. The situation is a bit more subtle for p odd but by making clever use of the Bockstein one can find for each element x in $H^*(BV; \mathbb{F}_p)$ an element θ in the Steenrod algebra such that θx is not nilpotent in the classical sense, and this suffices to show that any homomorphism from a nilpotent module N to $H^*(BV; \mathbb{F}_p)$ is trivial.

The converse is much harder and uses once more the Carlsson modules $K(i)$.   □

2.2. Quillen's F-isomorphism theorem from the point of view of unstable modules

Theorem 8 and 9 clearly imply the following result.

Theorem 10. *Let p be a prime. Then G is a Quillen group iff for every elementary abelian p-group V the map*

$$\text{Hom}_{\mathcal{U}_p}(H^*(BG; \mathbb{F}_p), H^*(BV; \mathbb{F}_p))$$
$$\longleftarrow \text{Hom}_{\mathcal{U}_p}(\lim_{A_p(G)^{op}} H^*(BE; \mathbb{F}_p), H^*(BV; \mathbb{F}_p))$$

induced by q_G is an isomorphism.   □

Now assume that the category $A_p(G)$ is equivalent to a finite category $A'_p(G)$. We remark that this is satisfied in many interesting cases: it is clearly satisfied if G is finite and it can be shown to hold if G is compact Lie, or more generally if G is a Quillen group for which $H^*(BG; \mathbb{F}_p)$ is a finitely generated $\mathbb{F}_p$-algebra. This includes many discrete groups, e.g. all $\mathcal{K}_1 \mathcal{F}$-groups in the sense of Definition 8 below, in particular all (S)-arithmetic groups, mapping class groups, automorphism groups of free groups, ...

Then the limit $\lim_{\mathcal{A}_p(G)^{op}} H^*(BE; \mathbb{F}_p)$ is isomorphic to the kernel of a map

$$\prod_E H^*(BE; \mathbb{F}_p) \longrightarrow \prod_{E' \longrightarrow E''} H^*(BE'; \mathbb{F}_p)$$

between finite products, indexed by the objects resp. morphisms in $\mathcal{A}'_p(G)$. Together with the injectivity of $H^*(BV; \mathbb{F}_p)$ this leads to the following iso-morphism

$$\mathrm{Hom}_{\mathcal{U}_p}(\lim_{\mathcal{A}_p(G)^{op}} H^*(BE; \mathbb{F}_p), H^*(BV; \mathbb{F}_p)) \cong$$
$$\mathrm{colim}_{\mathcal{A}_p(G)} \mathrm{Hom}_{\mathcal{U}_p}(H^*(BE; \mathbb{F}_p), H^*(BV; \mathbb{F}_p)) \ .$$

By a theorem of Adams, Gunawardena and Miller [AGM] (see also Corollary 17 below) the natural map

$$\mathbb{F}_p[\mathrm{Hom}(V, E)] \longrightarrow \mathrm{Hom}_{\mathcal{U}_p}(H^*(BE; \mathbb{F}_p), H^*(BV; \mathbb{F}_p))$$

is an isomorphism where for a set S the $\mathbb{F}_p$-vector space with S as a basis is denoted by $\mathbb{F}_p[S]$. Finally let $\mathrm{Rep}(V, G)$ denote the set of G-conjugacy classes of homomorphisms from V to G: then the maps $\mathrm{Hom}(V, E) \to \mathrm{Rep}(V, G), \alpha \mapsto i_E \alpha$ (with i_E denoting the inclusion of E into G) induce a bijection

$$\mathrm{colim}_{\mathcal{A}_p(G)} \mathrm{Hom}(V, E) \longrightarrow \mathrm{Rep}(V, G),$$

and hence also an isomorphism

$$\mathrm{colim}_{\mathcal{A}_p(G)} \mathbb{F}_p[\mathrm{Hom}(V, E)] \longrightarrow \mathbb{F}_p[\mathrm{Rep}(V, G)] \ .$$

In other words, we obtain the following characterization of Quillen groups.

Theorem 11. *Let p be a prime and let G be a group for which $\mathcal{A}_p(G)$ is equivalent to a finite category. Then G is a Quillen group iff the natural map*

$$\mathbb{F}_p[\mathrm{Rep}(V, G)] \longrightarrow \mathrm{Hom}_{\mathcal{U}_p}(H^*(BG; \mathbb{F}_p), H^*(BV; \mathbb{F}_p))$$

is an isomorphism for all elementary abelian p-groups V. □

Theorem 11 says in particular that the F-isomorphism type of $H^*(BG; \mathbb{F}_p)$ is completely captured by the sets $\mathrm{Rep}(V, G)$ resp. the vector spaces

$$\mathbb{F}_p[\mathrm{Rep}(V, G)] \quad \text{(with } V \text{ varying)},$$

at least if $\mathcal{A}_p(G)$ is finite up to equivalence. In particular, statements on $H^*(BG; \mathbb{F}_p)$ can often be tested by the corresponding statements on these sets. For more on this point of view we refer to [HLS1].

In the next section we will see how Lannes' theory of the T-functor can be used to compute, without using Quillen's theorem,

$$\mathrm{Hom}_{\mathcal{U}_p}(H^*(BG; \mathbb{F}_p), H^*(BV; \mathbb{F}_p)) \quad \text{as} \quad \mathbb{F}_p[\mathrm{Rep}(V, G)].$$

In fact, Lannes' results should be considered as a far reaching generalization of Theorem 11.

3. Lannes' T-functor and Lannes' generalization of Quillen's theorem

3.1. Review of Lannes' T-functor

The aim of this section is to introduce Lannes' T-functor and to review its main properties, at least as far as they are important for us. For details we refer to [Ln1], [Ln3] or [Sc].

Throughout this section p is a fixed prime. From now on we suppress the coefficients and write H^*BG instead of $H^*(BG; \mathbb{F}_p)$, and so on.

Given an elementary abelian p-group V, Lannes ([Ln1], [Ln3]) considered the functor $T_V : \mathcal{U} \longrightarrow \mathcal{U}$ characterized by the adjunction property

$$\mathrm{Hom}_{\mathcal{U}}(T_V M, N) \cong \mathrm{Hom}_{\mathcal{U}}(M, H^*BV \otimes N) \ .$$

One should view T_V as a "cohomological analogue" of the mapping space functor $X \mapsto \mathrm{map}(BV, X)$.

The existence of such a functor is quite formal: if T_V exists it is clearly right exact and preserves arbitrary direct sums, hence it is determined by its values on the free unstable modules $F(k)$ which are characterized by natural isomorphisms $\mathrm{Hom}_{\mathcal{U}}(F(k), N) \cong N^k$. The desired adjunction property of T_V forces $T_V F(n) \cong \bigoplus_k F(n-k)^{\oplus \dim H^k V}$. Conversely one can define T_V on free objects by this formula and use free presentations to define T_V on general objects.

Excercise 4: Show that $T_V B \cong B$ whenever B is a module which is bounded above, i.e. for which $B^n = 0$ if n is sufficiently large.

What makes these functors interesting is that they enjoy some magical properties which make them extremely useful. Here are the two basic properties of these functors. The first one is quite surprising, in particular in view of the analogy with the mapping space functor.

Theorem 12 [Ln1], [Ln3]. T_V *is an exact functor.*

Theorem 13 [Ln1], [Ln3]. T_V *commutes with tensor products, i.e. the product on H^*V induces a natural transformation $\tau_{M,N} : T_V(M \otimes N) \longrightarrow T_V M \otimes T_V N$ which is an isomorphism for all unstable modules M and N.*

More precisely if $\gamma_M : M \longrightarrow H^*BV \otimes T_V M$ denotes the adjoint of $id_{T_V M}$, then $\tau_{M,N}$ is the adjoint of the composition of the three maps

$$\gamma_M \otimes \gamma_N : M \otimes N \longrightarrow H^*BV \otimes T_V M \otimes H^*BV \otimes T_V N$$

$$H^*BV \otimes T_V M \otimes H^*BV \otimes T_V N \longrightarrow H^*BV \otimes H^*BV \otimes T_V M \otimes T_V N$$

$$\phi \otimes \ id \otimes id : H^*BV \otimes H^*BV \otimes T_V M \otimes T_V N \longrightarrow H^*BV \otimes T_V M \otimes T_V N$$

where the second map interchanges the second and third tensor factor, and ϕ is the product of the unstable algebra H^*BV.

Comments on the proofs of Theorem 12 and 13. Theorem 12 is really equivalent to the following result due to Lannes and Zarati. (Recall that $J(n)$ denotes the n-th Brown Gitler module representing the functor $M \longrightarrow (M^n)^*$.)

Theorem 12′ [LZ]. $H^*BV \otimes J(n)$ *is injective for each elementary abelian p-group V and each natural number n.*

Exactness of T_V reduces the proof of Theorem 13 to a verification that $\tau_{M,N}$ is an isomorphism for $M = F(m)$ and $N = F(n)$ and by the adjointness property of T_V this amounts to the computation of $\mathrm{Hom}_{\mathcal{U}}(F(m) \otimes F(n), H^*BV \otimes J(k))$ for all m, n and k. $\qquad\qquad\square$

The following two results are derived from Theorem 12 and Theorem 13 and are also of fundamental importance.

Theorem 14 [Ln1], [Ln3]. T_V *lifts to a functor from $\mathcal{K}$ to itself and continues to be adjoint to H^*BV, i.e. there are natural isomorphisms*

$$\mathrm{Hom}_{\mathcal{K}}(T_V K, L) \cong \mathrm{Hom}_{\mathcal{K}}(K, H^*BV \otimes L)$$

for all $K, L \in \mathcal{K}$.

(This is almost trivial given Theorem 13 except that one has to verify that on $T_V K$ the map $x \mapsto Sq^{|x|}x$ agrees with the squaring map if $p = 2$ resp. that $x \mapsto P^{\frac{|x|}{2}}x$ agrees with the p-th power map if p is odd.)

In order to state the next result in its proper generality we need to talk a little bit about profinite vector spaces and profinite sets.

We note that any unstable module is the colimit of its finitely generated submodules M_α, and this implies that $\mathrm{Hom}_{\mathcal{U}}(M, N)$ is the limit of the $\mathrm{Hom}_{\mathcal{U}}(M_\alpha, N)$. If N is of finite type (e.g. $N = H^*BV$) then $\mathrm{Hom}_{\mathcal{U}}(M_\alpha, N)$ is a finite dimensional $\mathbb{F}_p$-vector space for each α and hence $\mathrm{Hom}_{\mathcal{U}}(M, N)$ is a limit of finite $\mathbb{F}_p$-vector spaces, i.e. a profinite vector space. Similarly, $\mathrm{Hom}_{\mathcal{K}}(K, L)$ is a profinite set if L is of finite type. For a profinite $\mathbb{F}_p$-vector space E we denote the continuous dual (the space of continuous homomorphisms from E to $\mathbb{F}_p$) by E' and for a profinite set S we denote the set of all continuous functions from S to $\mathbb{F}_p$ by $\mathbb{F}_p^S$. We note that the category $\mathcal{E}_\infty$ of all $\mathbb{F}_p$-vector spaces is equivalent to the opposite of the category $\mathcal{PE}$ of all profinite $\mathbb{F}_p$-vector spaces and the equivalence is induced by taking the dual resp. the continuous dual.

Of course, if E is finite dimensional, or S is finite, the profinite topology is discrete and continuity is always guaranteed. For a first reading it will not do much harm to avoid the profinite structures by assuming appropriate finiteness conditions (cf. Corollary 16 below).

Theorem 15 [Ln1], [Ln3] (Linearization principle). *Let K be an unstable algebra. Then the natural restriction map*

$$T_V^0 K \cong \mathrm{Hom}_{\mathcal{U}}(K, H^* BV)' \cong \mathbb{F}_p^{\mathrm{Hom}_{\mathcal{K}}(K, H^* BV)}$$

is an isomorphism of p-Boolean algebras.

We recall that a p-Boolean algebra B is an $\mathbb{F}_p$-algebra with the property that $x^p = x$ for each $x \in B$. Clearly the (continuous) functions from any (profinite) set into $\mathbb{F}_p$ form a p-Boolean algebra, and the degree 0-part of any unstable algebra does as well (because the instability axioms say $x = Sq^0 x = x^2$ resp. $x = P^0 x = x^p$ if $|x| = 0$). Theorem 15 is an easy consequence of the structure theorem for p-Boolean algebras which states that a p-Boolean algebra B is isomorphic to the algebra of $\mathbb{F}_p$-valued continuous functions on the spectrum $\mathrm{spec}\, B$ of B, and $\mathrm{spec}\, B$ can be identified with the set of algebra homomorphisms from B to $\mathbb{F}_p$.

Corollary 16. *Let K be an unstable algebra. Then $\mathrm{Hom}_{\mathcal{U}}(K, H^* BV)$ is finite if and only if $\mathrm{Hom}_{\mathcal{K}}(K, H^* BV)$ is finite. In this case the natural homomorphism*

$$\mathbb{F}_p[\mathrm{Hom}_{\mathcal{K}}(K, H^* BV)] \longrightarrow \mathrm{Hom}_{\mathcal{U}}(K, H^* BV)$$

induced by the inclusion of $\mathrm{Hom}_{\mathcal{K}}(K, H^* BV)$ into $\mathrm{Hom}_{\mathcal{U}}(K, H^* BV)$ is an isomorphism. ☐

As a special case we obtain the theorem of Adams, Gunawardena and Miller (see the discussion before Theorem 11).

Corollary 17. *The natural map*

$$\mathbb{F}_p[\mathrm{Hom}(W, V)] \cong \mathbb{F}_p[\mathrm{Hom}_{\mathcal{K}}(H^* BV, H^* BW)] \longrightarrow \mathrm{Hom}_{\mathcal{U}}(H^* BV, H^* BV)$$

is an isomorphism. ☐

If you haven't already succeeded before, you should now be able to give the answer to Exercise 2b.

As a final application of Theorem 14 we list the following result.

Theorem 18. *Let W be an elementary abelian p-group and let $\alpha \in \mathrm{Hom}(V, W)$. Then the maps $V \times W \to W, (v, w) \mapsto \alpha(v) + w$ induce, via adjunction, maps of unstable algebras $T_V H^* BW \longrightarrow H^* BW$ such that the product of these maps gives an isomorphism of unstable algebras*

$$T_V H^* BW \longrightarrow \prod_{\alpha \in \mathrm{Hom}(V, W)} H^* BW \; .$$

Sketch of proof. There is a functor $U : \mathcal{U} \longrightarrow \mathcal{K}$ which is left adjoint to the forgetful functor. (If M is an unstable module then the symmetric algebra SM is an unstable module with an $\mathbb{F}_p$-algebra structure which satisfies the Cartan formula, but the formula $Sq^{|x|}x = x^2$ resp. $P^{\frac{|x|}{2}}x = x^p$ does not hold. UM is the quotient of SM by the ideal generated by the elements $Sq^{|x|}x - x^2$ resp. $P^{\frac{|x|}{2}}x - x^p$.)

One can easily check that $H^*B\mathbb{Z}/p \cong UF(1)$ where $F(1)$ is the free unstable module on a generator in degree 1. Next it is clear that T_V commutes with the functor U and this implies the computation of $T_V H^*B\mathbb{Z}/p$, and finally of $T_V H^*BW$ because T_V commutes with tensor products.   □

3.2. Lannes' generalization of Quillen's theorem

We go back and have another look at Theorem 11. Taking $N = \mathbb{F}_p$ (considered as an unstable module concentrated in degree 0) in the adjunction formula for T_V we see that Theorem 11 is nothing but the computation of $(T_V^0 H^*BG)^*$ with T_V^0 denoting the degree 0 part of T_V. In [Ln2] Lannes computes all of $T_V H^*BG$ and this computation should be considered as a far reaching generalization of Quillen's theorem. To describe his result we need to introduce some more notation.

For a representation $\rho \in \mathrm{Rep}(V, G)$ we choose a representative and by abuse of notation we will denote this representative again by ρ. Then we consider the homomorphism

$$c_\rho : V \times C_G(\rho) \longrightarrow G, \quad (v, g) \mapsto \rho(v)g .$$

Here $C_G(\rho)$ stands for the centralizer of $\mathrm{Im}\,\rho$ in G. We obtain an induced map $c_\rho^* : H^*BG \longrightarrow H^*BV \otimes H^*BC_G(\rho)$ and an adjoint map $ad(c_\rho^*) : T_V H^*BG \longrightarrow H^*BC_G(\rho)$.

Theorem 19 [Ln2], [Ln3]. *Let G be a finite group and V be an elementary abelian p-group. Pick a representative for each element ρ in $\mathrm{Rep}(V, G)$. Then the resulting map*

$$l_G^V : T_V H^*BG \longrightarrow \prod_{\rho \in \mathrm{Rep}(V,G)} H^*BC_G(\rho)$$

with components $ad(c_\rho^*)$ is an isomorphism in $\mathcal{K}$.

As in the case of Theorem 5 this result also holds for other classes of groups, in particular it holds for all classes of groups listed after Theorem 5. In the case of groups of finite virtual mod-p cohomological dimension and in the case of compact Lie groups this was shown in [Ln2] and for profinite groups in [H4]. Still another class was considered in the appendix of [H2].

Definition 6. *We call a topological group G a Lannes group at the prime p if the map l_G^V is an isomorphism for all elementary abelian p-groups V (with modifications as in Section 1 if G is a profinite group).*

Clearly, a Lannes group for which $\mathcal{A}_p(G)$ is equivalent to a finite category is also a Quillen group. The class of Lannes groups is, just as the class of Quillen groups, not well understood, and it would be nice to have a characterization of this class of groups in more group theoretical terms.

Theorem 19 has various applications to the study of group cohomology which do not involve the Steenrod algebra and which refine Quillen's results (cf. Sections 4, 5 and 6 below). Philosophically, Theorem 19 emphasizes the role of centralizers of elementary abelian p-subgroups as a second deeper level of information on H^*BG, the first level being given by the category of elementary abelian p-subgroups of G. We will elaborate on this philosophy in the next two sections.

Outline of a proof of Theorem 19 (for G compact Lie). The first step in the proof is to generalize the result to be proved. So we consider G-CW-complexes X and the functor $L_1 = L_1^G$ from the category of G-CW-complexes to graded $\mathbb{F}_p$-vector spaces, say, which assigns to X the graded vector space $T_V H_G^* X$ where $H_G^* X$ denotes the equivariant cohomology $H_G^*(X)$, i.e. the cohomology of the space $EG \times_G X$. (Here EG denotes as usual the total space of the universal principal G-bundle over BG; so if G is discrete, EG is just the universal cover of BG.) We will usually suppress G from the notation if the group is known from the context. Note that if X is a point then $L_1(X)$ is just the source of the map l_G^V.

If $\rho : V \longrightarrow G$ is a homomorphism let X^ρ denote the subspace fixed by the image of ρ, i.e. $X^\rho := \{x \in X | \rho(v)x = x \ \forall \ v \in V\}$. The centralizer $C_G(\rho)$ acts in a natural way on X^ρ so we can consider $H_{C_G(\rho)}^*(X^\rho)$ and the functor $L_2 = L_2^G$ (again G will usually be suppressed) from the category of G-CW-complexes to graded $\mathbb{F}_p$-vector spaces, say, which assigns to X the product $\prod_{\rho \in \text{Rep}(V,G)} H_{C_G(\rho)}^*(X^\rho)$. If X is a point this is just the target of the map l_G^V.

Next one defines a natural transformation $l_{G,X}^V : L_1(X) \longrightarrow L_2(X)$ which for X a point specializes to l_G^V; its ρ-th component (for $\rho \in \text{Rep}(V,G)$) is adjoint to a map $H_G^*(X) \longrightarrow H^*BV \otimes H_{C_G(\rho)}^*(X^\rho)$ which is defined as follows: the inclusion $X^\rho \subset X$ and the homomorphism $V \times C_G(\rho) \to G, (v,g) \mapsto \rho(v)g$ induce a map $BV \times (EC_G(\rho) \times_{C_G(\rho)} X^\rho) \longrightarrow EG \times_G X$ and the induced map in cohomology is the required map. Now the generalization of Theorem 19 reads as follows:

Theorem 19′. *Suppose G is a compact Lie group. Then the map $l_{G,X}^V : L_1(X) \longrightarrow L_2(X)$ is an isomorphism for each finite G-CW-complex X.*

Outline of the proof of Theorem 19′. The proof uses the same strategy as Quillen's original proof of Theorem 5. The functors L_1 and L_2 are G-homotopy invariant and associate to a homotopy pushout Mayer-Vietoris sequences (for L_1 this uses the exactness of T_V). They both turn finite coproducts into finite products and therefore $l^V_{G,X} : L_1(X) \longrightarrow L_2(X)$ is an isomorphism if $l^V_{G,G/K} : L_1(G/K) \longrightarrow L_2(G/K)$ is an isomorphism for each G-orbit G/K which appears in X.

The crucial step is now to reduce to the case where all isotropy subgroups are elementary abelian (cf. Section 6 in [Q1]). To this end we consider an embedding of G into a unitary group and consider the resulting G-space $F := U(n)/S$ where S is the elementary abelian subgroup of $U(n)$ generated by all diagonal matrices of order p. Then one considers the following coequalizer diagram of G-spaces

$$X \times F \times F \overset{p_2}{\underset{p_1}{\rightrightarrows}} X \times F \overset{p}{\longrightarrow} X$$

with $p_1(x,f,f') = (x,f)$, $p_2(x,f,f') = (x,f')$ and $p(x,f) = x$. The crucial observation by Quillen was that after applying cohomology one obtains an equalizer diagram of $\mathbb{F}_p$-algebras (even unstable algebras)

$$(*) \qquad H^*_G(X) \overset{p}{\longrightarrow} H^*_G(X \times F) \overset{p^*_2}{\underset{p^*_1}{\rightrightarrows}} H^*_G(X \times F \times F)$$

The point of all this is that the isotropy subgroups for the G-action on F are all elementary abelian (they are conjugate in $U(n)$ to subgroups of S!) and hence the isotropy groups for the G-action on $X \times F$ and $X \times F \times F$ are also elementary abelian.

Now exactness of T_V implies that we have an equalizer diagram

$$L_1(X) \overset{p}{\longrightarrow} L_1(X \times F) \overset{p^*_2}{\underset{p^*_1}{\rightrightarrows}} L_1(X \times F \times F)$$

and a modification of Quillen's argument shows that

$$L_2(X) \overset{p}{\longrightarrow} L_2(X \times F) \overset{p^*_2}{\underset{p^*_1}{\rightrightarrows}} L_2(X \times F \times F)$$

is also an equalizer diagram (in fact a product of equalizer diagrams, one for each $\rho \in \mathrm{Rep}(V, G)$). Therefore it suffices to show the theorem for the G-spaces $X \times F$ resp. $X \times F \times F$, i.e. it is now enough to check the theorem for G-orbits G/E where E is an elementary abelian p-subgroup of G.

Next one checks that for $i = 1, 2$ there are natural "Shapiro type" isomorphisms $L^G_i(G/E) \cong L^E_i(pt)$ where pt denotes a one point space. For $i = 1$ this is an immediate consequence of the usual Shapiro isomorphism $H^*_G(G/E) \cong H^*_E(pt)$; for $i = 2$ it is an interesting exercise (see also Lemma 2.8 in [H3]). Furthermore, one verifies that with respect to these isomorphisms the map $l^V_{G,G/E}$ identifies with $l^E_{V,pt}$. Therefore it suffices to verify Theorem 19 in the case $G = E$ and this is guaranteed by Theorem 18. $\square$

3.3. An outline of a proof of Quillen's Theorem

The proof of Theorem 19 given above suggests the following strategy for a proof of Quillen's original theorem (for compact Lie groups) which is independent of the theory of unstable modules, and in fact, is nothing but a streamlined version of Quillen's original proof.

First one looks for a suitable generalization of Theorem 5. For this we need to introduce certain functors from $\mathcal{A}_p(G)^{op}$ to the category $\mathcal{E}ns$ of sets. The category of such functors will also be denoted by $\mathcal{A}_p(G) - \mathcal{E}ns$; its objects will be called $\mathcal{A}_p(G)$-sets.

The functor $E \mapsto H^*BE$ will be simply denoted by H^*.

If X is a G-space then the centralizer $C_G(E)$ acts on the E-fixed points X^E and, if π_0 denotes the set of path components, we can form the set $\pi_0(X^E)/C_G(E)$.

Then it is easy to check that the assignment $E \mapsto \pi_0(X^E)/C_G(E)$ extends canonically to a functor $F_X : \mathcal{A}_p(G)^{op} \to \mathcal{E}ns$. If we want to emphasize the group as well we will denote this functor by $F_{G,X}$.

For a G-space X we obtain a map $q_{G,X}$ from $H^*_G X$ to the set of natural transformations $\mathrm{Hom}_{\mathcal{A}_p(G)-\mathcal{E}ns}(F_X, H^*)$ of $\mathcal{A}_p(G)$-sets as follows: For $c \in \pi_0(X^E)$ we choose a point x_c in the component of c. The inclusions $\{x_c\} \subset X^E \subset X$ on the space level and $E \subset G$ on the group level induce a map $c^* : H^*_G X \longrightarrow H^*BE$ which one can easily check to be independent of the choice of x_c. If $u \in H^*_G X$ then the E-th component of $q_{G,X}(u)$ maps c to $c^*(u)$; it is straightforward to check that $q_{G,X}(u)$ is a natural transformation from F_X to H^*. Furthermore the target of $q_{G,X}$ is a graded $\mathbb{F}_p$-algebra (because H^* takes values in graded $\mathbb{F}_p$-algebras) and the map $q_{G,X}$ is a homomorphism of graded $\mathbb{F}_p$-algebras.

Here is the generalization of Theorem 5.

Theorem 5'. *Assume G is a compact Lie group and X is a finite G-CW-complex. Then the natural map*

$$q_{G,X} : H^*_G X \longrightarrow \mathrm{Hom}_{\mathcal{A}_p(G)-\mathcal{E}ns}(F_X, H^*)$$

is an F-isomorphism.

If X is a point then the functor F_X is the constant functor with value a point; in this case the target is nothing but $\lim_{\mathcal{A}_p(G)^{op}} H^*$ and $q_{G,X}$ is just Quillen's map q_G.

Sketch of proof of Theorem 5′. Abbreviate the source of $q_{G,X}$ by $Q_1(X)$ and its target by $Q_2(X)$. Then both Q_1 and Q_2 are functors from G-spaces to graded $\mathbb{F}_p$-algebras which are invariant with respect to G-homotopy equivalences and turn finite coproducts into products. Furthermore Q_2 turns a homotopy pushout into a pullback of graded $\mathbb{F}_p$-algebras; Q_1 associates to a homotopy pushout a Mayer-Vietoris sequence and this means (cf. Proposition 3.2 in [Q1]) that up to F-isomorphism it also turns a homotopy pushout into a pullback. (Note that it is precisely at this point where we have to give up the hope for a genuine isomorphism in Quillen's Theorem.)

Consequently, we may concentrate on the case of an orbit $X = G/K$. Using Quillen's trick (cf. Section 6 in [Q1]) of embedding G into a unitary group $U(n)$ and working with the projection $X \times F \longrightarrow X$, one reduces the case of a general G-space to the case of an orbit $X = G/E$ with E elementary abelian.

Then one identifies $q_{G,G/E}$ with $q_{E,pt}$ as above and finally one checks that the map $q_{E,pt}$ is a (genuine!) isomorphism. □

4. Approximations up to higher nilpotency

4.1. Localizations in abelian categories

We consider Quillen's Theorem as a first order approximation to H^*BG and we will discuss other approximations in this and the following section. The philosophy behind these approximations is best understood from the point of view of localizations in abelian categories. We try to explain this is in a non-technical way and therefore we deliberately remain a bit vague. For more details we refer to Gabriel [Ga].

So suppose $\mathcal{C}$ is a (small) abelian category and $\mathcal{I}$ is a set of injective objects. Consider the following subcategory $\mathcal{N}(\mathcal{I})$ of $\mathcal{C}$: An object N belongs to $\mathcal{N}(\mathcal{I})$ iff $\mathrm{Hom}_{\mathcal{C}}(N, I) = 0$ for all $I \in \mathcal{I}$. Then $\mathcal{N}(\mathcal{I})$ is a *Serre subcategory*, i.e. if

$$0 \longrightarrow M_1 \longrightarrow M_2 \longrightarrow M_3 \longrightarrow 0$$

is an exact sequence in the category $\mathcal{C}$, then M_2 is in $\mathcal{N}(\mathcal{I})$ if and only if M_1 and M_3 are.

In such a situation one might want to systematically ignore objects in $\mathcal{N}(\mathcal{I})$, e.g. one might like to regard those morphisms as isomorphisms for which the kernel and cokernel are in $\mathcal{N}(\mathcal{I})$. Technically this can be achieved by passing to the quotient category $\mathcal{C}/\mathcal{N}(\mathcal{I})$; this is an abelian category which has the same objects as $\mathcal{C}$ but one changes the morphisms in such a way that the objects in $\mathcal{N}(\mathcal{I})$ become isomorphic to the trivial object, but no others. (Can you make this precise?)

Often there is an alternative to passing to this quotient category, called *localization away from* $\mathcal{N}(\mathcal{I})$. The category $\mathcal{N}(\mathcal{I})$ is called a localizing subcategory if there is a functor $L : \mathcal{C} \longrightarrow \mathcal{C}$ and a natural transformation $\lambda : id_{\mathcal{C}} \longrightarrow L$ such that the following statements hold for every object C of $\mathcal{C}$:

- The kernel and cokernel of the map $\lambda : C \longrightarrow LC$ are in $\mathcal{N}(\mathcal{I})$.
- The object LC is $\mathcal{N}(\mathcal{I})$-closed, i.e. whenever $f : A \longrightarrow B$ is a morphism in $\mathcal{C}$ with $\mathrm{Ker}\, f$ and $\mathrm{Coker}\, f$ in $\mathcal{N}(\mathcal{I})$, then the induced map $\mathrm{Hom}_{\mathcal{U}}(f, C)$ is an isomorphism.

If localization away from $\mathcal{N}(\mathcal{I})$ exists it is clearly unique up to natural equivalence and it picks for each object C a particularly nice object, namely LC, among all those objects which in the quotient category become isomorphic to C.

Injectivity of $I \in \mathcal{I}$ implies that λ induces an isomorphism

$$\mathrm{Hom}_{\mathcal{C}}(LC, I) \cong \mathrm{Hom}_{\mathcal{C}}(C, I).$$

This shows that LC depends on the functor $F_C : \mathcal{I} \to \mathcal{A}b, I \mapsto \mathrm{Hom}_{\mathcal{C}}(C, I)$ (where $\mathcal{I}$ is considered as the full subcategory with object set $\mathcal{I}$ and $\mathcal{A}b$ denotes the category of abelian groups). In fact, in favourable cases LC can even be reconstructed from this functor. For example, if $\mathcal{I}$ is a finite category (i.e. the set of objects and the set of morphisms is finite) and the functor F_C takes values in finite abelian groups, then LC can be described as follows:

Let $\mathrm{Hom}_+(F_C(I), I)$ denote additive maps from the abelian group $F_C(I)$ to I. (In case $F_C(I)$ is finite one can make sense out of this in an arbitrary abelian category. Can you explain how?) There is a natural "evaluation map" $\varepsilon : C \longrightarrow \prod_{I \in \mathcal{I}} \mathrm{Hom}_+(F_C(I), I)$ and it is easy to check that the kernel of this map is the largest subobject of C which is in $\mathcal{N}(\mathcal{I})$. Furthermore LC can be described as a subobject of the product and ε induces the localization map λ. More precisely, the map ε is easily seen to be equalized by the two natural maps

$$\prod_I \mathrm{Hom}_+(F_C(I), I) \rightrightarrows \prod_{I_1 \to I_2} \mathrm{Hom}_+(F_C(I_1), I_2)$$

where the products extend over all objects resp. morphisms in $\mathcal{I}$. Now it is not hard to check that the equalizer of these two maps is $\mathcal{N}(\mathcal{I})$-closed and the map from C to the equalizer qualifies as localization away from $\mathcal{N}(\mathcal{I})$.

As long as the products above make sense the map from C to the equalizer is always a good candidate for the localization, even if the finiteness assumptions on $\mathcal{I}$ and F_C are not satisfied. In particular the equalizer is always $\mathcal{N}(\mathcal{I})$-closed (cf. Exercise 5 below). However, it may be hard to compute the group of homomorphisms out of the equalizer (because maps out of an arbitrary product are usually difficult to understand), and hence it may be hard to verify that λ has kernel and cokernel in $\mathcal{N}(\mathcal{I})$. Nevertheless the philosophy still works in some interesting cases where these finiteness assumptions are not satisfied.

Localization away from $\mathcal{N}(\mathcal{I})$ is particularly interesting if the category $\mathcal{N}(\mathcal{I})$ has interesting properties and if LC (i.e. the functor F_C) is computable. In the case of the category of unstable modules (and variations thereof) the functor F_{H^*BG} is often "computable" via Lannes' Theorem, and hence LH^*BG is often accessible. In fact, at least philosophically the results in Section 4.2 and Section 5 should be looked at from this point of view.

Example. Consider the category $\mathcal{U}$. Pick a skeleton $\mathcal{E}_s$ of the category $\mathcal{E}$ of elementary abelian p-groups and consider $\mathcal{I} = \{H^*BV | V \in \mathcal{E}_s\}$. By Theorem 9 we see that in this case $\mathcal{N}(\mathcal{I})$ is the full subcategory $\mathcal{N}il$ of nilpotent modules. Localization away from $\mathcal{N}il$ has been studied in [HLS1], [HLS2].

Excercise 5: Assume $\mathcal{C}$ is an abelian category and $\mathcal{I}$ is a set of injectives in $\mathcal{C}$. Show that

a) I is $\mathcal{N}(\mathcal{I})$-closed for every $I \in \mathcal{I}$.

b) A product of $\mathcal{N}(\mathcal{I})$-closed objects is $\mathcal{N}(\mathcal{I})$-closed. (This holds even for arbitrary products as long as they exist).

c) If M_1 and M_2 are $\mathcal{N}(\mathcal{I})$-closed and if M is the kernel of a map $M_1 \longrightarrow M_2$ then M is $\mathcal{N}(\mathcal{I})$-closed.

d) Let $\mathcal{C} = \mathcal{U}$, $\mathcal{N}(\mathcal{I}) = \mathcal{N}il$ and assume G is a Quillen group. Conclude that up to natural equivalence Quillen's map q_G qualifies as localization away from $\mathcal{N}il$.

4.2. Localization away from n-nilpotent modules

In this section we will explain some of the main results of [HLS3].

Given the discussion above it is natural to consider localization functors on $\mathcal{U}$ based on other sets of injectives. Let $n \in \mathbb{N}$, $n > 0$. In this section we will consider the set $\mathcal{I}_n := \{H^*BV \otimes J(k) | V \in \mathcal{E}_s, k = 0, \dots, n-1\}$. The values $F_M(H^*BV \otimes J(k))$ of the functor F_M are in this case given by the dual of the vector space $T_V^k(M)$, in particular they have a beautiful interpretation if G is a Lannes group and $M = H^*BG$. Moreover the category $\mathcal{N}(\mathcal{I}_n)$ also has a beautiful characterization given as follows.

Definition 7 – Proposition 20. *Let n be a natural number. An unstable module N is called n-nilpotent if the following equivalent conditions hold*

(1) *N is in $\mathcal{N}(\mathcal{I}_n)$.*

(2) *$T_V M$ is $(n-1)$-connected, i.e. $T_V^k M = 0$ for $k < n$.*

(3) *Every finitely generated submodule of M has a finite filtration by unstable modules such that the successive quotients in this filtration are n-fold suspensions of unstable modules.*

The full subcategory of all n-nilpotent modules is usually denoted $\mathcal{N}il_n$ instead of $\mathcal{N}(\mathcal{I}_n)$.

It is obvious that (1) and (2) are equivalent by the defining property of the Brown Gitler modules $J(n)$. For the equivalence of (1) and (3) we refer to [Sc].

Note that Theorem 9 says that an unstable module is 1-nilpotent in the sense of Definition 7 if and only if it is nilpotent in the sense of Definition 5.

The third characterization of n-nilpotent modules shows that $\mathcal{N}il_n$ is the smallest Serre subcategory of $\mathcal{U}$ which contains all n-fold suspensions and which is closed with respect to filtered colimits. There is a characterization of n-nilpotent modules in terms of Steenrod operations (cf. [H6]) but the characterization given here is usually more useful and in any case more conceptual. We emphasize that the concept relies crucially on the presence of Steenrod operations and the instability condition (it is the instability axiom which prevents a general unstable module from being a suspension!).

The subcategories $\mathcal{N}il_n$ turn out to be localizing subcategories. The corresponding localization functors and natural transformations will be denoted by L_n and λ_n. We will now describe the localization of H^*BG away from $\mathcal{N}il_n$. The point of view in the following discussion is a bit different from the discussion in Section 4.1 in so far as we do not construct L_n and λ_n from the corresponding functor

$$F_{H^*BG} : \mathcal{I}_n \to \mathcal{A}b, I \mapsto \mathrm{Hom}_{\mathcal{U}}(H^*BG, I)$$

precisely as suggested above. Instead, we process the information carried by this functor in a way which is more adequate from the point of view of group cohomology. The resulting description is equivalent to the one suggested in Section 4.1 (even though $\mathcal{I}_n$ does not satisfy the strong finiteness assumptions), but it is fair to say that the translation from one description to the other is subtle.

As in Section 4.1 we start by considering the following natural map which is induced by evaluation:

$$\varepsilon_n : M \longrightarrow \prod_{(V,k<n)} \mathrm{Hom}_+(\mathrm{Hom}_{\mathcal{U}}(M, H^*BV \otimes J(k)), H^*BV \otimes J(k)) \ .$$

Here the product ranges over all elementary abelian p-groups V and all $k < n$. The injectivity of the objects $H^*BV \otimes J(k)$ implies that the kernel K of this map satisfies property (1) of Definition 7, hence K is n-nilpotent, and in fact,

K is the largest n-nilpotent submodule of M. By the defining property of T_V it is clear that ε_n factors through the map $M \longrightarrow \prod_V H^*BV \otimes T_V M$ whose components are the adjoints of the identity of $T_V M$, and even through

$$M \longrightarrow \prod_V H^*BV \otimes (T_V M)^{<n}$$

where $(-)^{<n}$ denotes the quotient by the submodule generated by elements of degree at least n. We denote this last map by $\overline{\lambda}_n$. Furthermore, it is clear that the obvious map

$$\prod_V H^*BV \otimes (T_V M)^{<n}$$
$$\longrightarrow \prod_{(V,k<n)} \mathrm{Hom}_+(\mathrm{Hom}_{\mathcal{U}}(M, H^*BV \otimes J(k)), H^*BV \otimes J(k))$$

is injective, hence the kernel of ε_n is equal to the kernel of $\overline{\lambda}_n$.

Now consider the case $M = H^*BG$ where G is a Lannes group. By Theorem 19 the map $\overline{\lambda}_n$ takes the following form

$$\overline{\lambda}_n : H^*BG \longrightarrow \prod_V \prod_{\mathrm{Rep}(V,G)} H^*BV \otimes (H^*BC_G(\rho))^{<n} .$$

The components of this map are induced by the maps $c_\rho : V \times C_G(\rho) \to G$ and these can be factored as $V \times C_G(\rho) \to \mathrm{Im}\,\rho \times C_G(\rho) \to G$ with

$$H^*B\,\mathrm{Im}\,\rho \otimes H^*BC_G(\rho) \to H^*BV \otimes H^*BC_G(\rho)$$

injective. Therefore we have proved the following result.

Theorem 21 [HLS3]. *Let p be a prime and G be a Lannes group. Then the kernel of the map*

$$\widetilde{\lambda}_n : H^*BG \longrightarrow \prod_{E \in \mathcal{A}_p(G)} H^*BE \otimes (H^*BC_G(E))^{<n}$$

whose components are induced by the maps $c_E : E \times C_G(E) \to G, (e,g) \mapsto eg$ is the largest n-nilpotent submodule of H^*BG. □

Note that if $n = 1$ the map $\widetilde{\lambda}_n$ agrees with the map $H^*BG \longrightarrow \prod_E H^*BE$ considered in Quillen's theorem. The localization $L_n H^*BG$ can now, just as in the case $n = 1$, be described by compatibility conditions on elements in the target of $\widetilde{\lambda}_n$.

In case $n = 1$ the compatibility conditions are completely determined by the category $\mathcal{A}_p(G)$. In the general case there are additional conditions; roughly speaking they arise from varying the groups E and the fact that the two homomorphisms

$$E \times E \times C_G(E) \longrightarrow E \times C_G(E)$$

given by
$$(e_1, e_2, g) \mapsto (e_1 e_2, g) \quad \text{resp.} \quad (e_1, e_2, g) \mapsto (e_1, e_2 g)$$
are equalized by the homomorphism
$$E \times C_G(E) \to C_G(E), \ (e, g) \mapsto eg \ .$$
(We will denote the former two homomorphisms by $m_E \times id$ resp. $id \times c_E$ where we have taken the liberty to use the symbol c_E again although here its target need not agree with G.) More generally (and this includes the phenomenon of varying the groups), if $\alpha : E_1 \to E_2, e \mapsto heh^{-1}$ is a morphism in the Quillen category, then we have two homomorphisms
$$E_1 \times E_1 \times C_G(E_2) \to E_1 \times C_G(E_1), (e_1, e_2, g) \mapsto (e_1 e_2, h^{-1} g h)$$
$$E_1 \times E_1 \times C_G(E_2) \to E_2 \times C_G(E_2), (e_1, e_2, g) \mapsto (h e_1 h^{-1}, h e_2 h^{-1} g)$$
whose compositions with the maps
$$c_{E_i} : E_i \times C_G(E_i) \to G, \ (e, g) \mapsto eg, \ i = 1, 2$$
agree up to conjugacy in G. Hence the two induced compositions (for $i = 1, 2$)
$$H^* BG \longrightarrow H^* BE_i \otimes (H^* BC_G(E_i))^{<n} \to H^* BE_1 \otimes (H^* BE_1 \otimes H^* BC_G(E_2))^{<n}$$
coincide.

Theorem 22 [HLS3]. *Let p be a prime and G be a Lannes group for which $\mathcal{A}_p(G)$ is equivalent to a finite category. The map $\widetilde{\lambda_n}$ induces a map from $H^* BG$ to the equalizer*

$$Eq : \prod_E H^* BE \otimes (H^* BC_G(E))^{<n}$$

$$\rightrightarrows \prod_{E_1 \to E_2} H^* BE_1 \otimes (H^* BE_1 \otimes H^* BC_G(E_2))^{<n}$$

and this induced map can be identified with λ_n, the localization away from $\mathcal{N}il_n$. (The products in this diagram extend over the objects resp. the morphisms in Quillen's category.)

We repeat (see Section 2.2) that the finiteness condition on the category is implied by the assumption that $H^* BG$ is a Lannes group with $H^* BG$ a finitely generated $\mathbb{F}_p$-algebra. Roughly speaking, Theorem 22 says that $L_n H^* BG$ is determined by $\mathcal{A}_p(G)$ and the mod-p cohomology of the groups $C_G(E)$ in degrees less than n.

Outline of the proof. There are two things to be checked: that the equalizer is $\mathcal{N}il_n$-closed, and that the map induced by λ_n has kernel and cokernel in $\mathcal{N}il_n$.

In fact, it is not hard to show that $H^* V \otimes F$ is $\mathcal{N}il_n$-closed whenever F is an unstable module which is concentrated in degrees less than n. Then Exercise 5 allows to conclude that the equalizer is $\mathcal{N}il_n$-closed.

To show that $\operatorname{Ker}\lambda_n$ and $\operatorname{Coker}\lambda_n$ are in $\mathcal{N}il_n$ it suffices to show that $\operatorname{Hom}_{\mathcal{U}}(\lambda_n, I)$ is an isomorphism for all $I \in \mathcal{I}_n$, in other words that $T_V(\lambda_n)$ is an isomorphism in degrees $< n$ for all elementary abelian V. Because of the magical properties of Lannes' T-functor this turns out to be manageable. $\square$

The proof that we just outlined is much more direct than the proof given in [HLS3] where (a version) of Theorem 22 was presented as a special case of the general study of localizations away from $\mathcal{N}il_n$. In fact, the determined reader should be able to supply the details of the proof. All he needs to know about Lannes' functor has been provided in Section 3.

In the following special case the compatibility conditions simplify drastically.

Corollary 23 [HLS3]. *Let G be a Lannes group for which $\mathcal{A}_p(G)$ is equivalent to a finite category. Assume that all elements of G of order p are central (and hence generate a central elementary abelian p-subgroup D). Then the map c_D induces a homomorphism of unstable algebras*

$$H^* BG \longrightarrow Eq : H^* BD \otimes (H^* BG)^{<n} \rightrightarrows H^* BD \otimes (H^* BD \otimes H^* BG)^{<n}$$

which can be identified with localization away from $\mathcal{N}il_n$. (Again Eq stands for equalizer and the two maps in the equalizer diagram are induced by the maps $m_D \times id$ resp. $id \times c_D$.)

Excercise 6: Derive Corollary 23 from Theorem 22.

One of the interests in these higher localizations is that they give a sequence of approximations which become better and better as n grows. In fact, the following result shows that this sequence stabilizes to a genuine isomorphism after a finite number of steps.

Theorem 24 [HLS3]. *Assume G is a Lannes group such that $H^* BG$ is finitely generated as an $\mathbb{F}_p$-algebra.*

a) *There exist natural numbers $d_0(G)$ and $d_1(G)$ such that λ_n (and hence $\widetilde{\lambda_n}$) is a monomorphism iff $n > d_0(G)$ and λ_n is an isomorphism iff $n > d_1(G)$.*

b) *Let G be a finite group or a compact Lie group and let U be a unitary group such that G embeds into U. Then we have the following inequalities:*

   - $d_0(G) \le \dim U - \dim G$,
   - $d_1(G) \le 2\dim U - \dim G$.
     (Here $\dim G$ etc. refer to the dimension of G considered as a manifold.)

Comment on the proof. Part (a) is a general fact about localizations of unstable modules which happen to be unstable algebras and are finitely generated as $\mathbb{F}_p$-algebras.

For the proof of Part (b) one uses once more Quillen's trick (i.e. the equalizer diagram (*) from the proof of Theorem 19'). This reduces the problem to finding bounds for d_0 and d_1 in the cases

$$H_G^*(U(n)/S) \cong H_S^*(G\backslash U(n))$$

and

$$H_G^*(U(n)/S \times U(n)/S) \cong H_{S \times S}^*(G\backslash U(n) \times U(n)) \ .$$

Therefore one is lead to study d_0 and d_1 in the case of equivariant cohomology of an elementary abelian group acting on a smooth manifold. This case can be handled by using a result of Duflot on the structure of such equivariant cohomology rings [Du1].   □

So the invariants $d_0(G)$ and $d_1(G)$ are not merely theoretical constants but they can be effectively bounded above. The bounds given in the theorem are very crude in general. For refinements we refer the reader to sections II.2 and II.3 of [HLS3]. We remark that the precise computation of the invariants $d_0(G)$ and $d_1(G)$ appears to be a difficult problem in general, and further investigation of these invariants should be worthwile.

There are also bounds for many discrete groups based on extensions of Theorem 24 to equivariant cohomology (see [HLS3]) but the bounds are less simple to state.

Excercise 7: Use the known computations of H^*BG (cf. [AM]) together with the description of λ_n given in Corollary 23 to determine the invariants d_0 and d_1 for the groups $\mathbb{Z}/p^n$ and S^1 at any prime, and for Q_8 and S_3 at the prime 2.

In principle $L_n H^*BG$ (and thus H^*BG) can be computed from knowing the category $\mathcal{A}_p(G)$ and the low dimensional cohomology of the spaces $BC_G(E)$. The examples in Exercise 7 suggest that this may not be a particularly efficient way to compute the cohomology of a finite group, at least not in cases which are accessible by traditional methods. The situation might change, however, if the groups get sufficiently complicated. In any case, if G is a finite group, there is a precise algorithm starting from a well defined finite set of data. This algorithm has not been seriously tested yet but one can hope that with increasing computer power this may become a feasable method for computing the cohomology of a finite group.

The invariants d_0 and d_1 behave well with respect to certain group theoretic constructions like the wreath product construction. This can sometimes be used to keep the bounds for these invariants much smaller than those given by Theorem 24. For example, in [HLS3] we show that if $p = 2$ and if q is a prime power with $q \equiv 3 \bmod 4$ then $d_0(GL(n, \mathbb{F}_q)) = 0$, $d_1(GL(n, \mathbb{F}_q)) \leq 2$. From this we were able to recover Quillen's computation of the mod-2 cohomology of the groups $GL(n, \mathbb{F}_q)$ [Q2].

As an immediate further consequence of Theorem 21 and Theorem 24 we get some information on the nilpotency-height of the ideal of nilpotent elements in H^*BG (in terms of the invariant $d_0(G)$). For $p = 2$ this ideal agrees with the kernel $N(G)$ of the restriction map $H^*BG \longrightarrow \prod_{E \in \mathcal{A}_2(G)} H^*BE$, for odd primes it is closely related to $N(G)$.

Corollary 25 [HLS3]. *Let G be a Lannes group for which H^*BG is a finitely generated algebra. Then the n-th power of the ideal $N(G)$ is trivial if $n > d_0(G)$.*

Proof. If x is in $N(G)$ then x restricts trivially to H^*BE for all $E \in \mathcal{A}_p(G)$ and hence, if $\widetilde{H}^*$ denotes the ideal of elements of positive degree, $c_E^*(x) \in H^*BE \otimes (\widetilde{H}^*BC_G(E))^{<n}$. Consequently, if $x_1, ..., x_n$ are in $N(G)$ then $c_E^*(x_1...x_n) = 0$ and we are done by Theorem 24. $\qquad\qquad\qquad\qquad\qquad\qquad\qquad\qquad\qquad\qquad\square$

It is not hard to produce (say by considering cartesian products) for each $n \in \mathbb{N}$ a finite p-group F_n such that $N(F_n)^n \neq 0$. For $p = 2$ even more is true as we have already remarked in Section 1: by ([AC], [IK]) there exist finite 2-groups F_n with elements $t_n \in H^*(BF_n; \mathbb{F}_2)$ such that $t_n^n \neq 0$, $t_n^{n+1} = 0$.

5. Commutative algebra of unstable H^*BG-modules; approximations of H^*BG up to torsion modules

Throughout this section p is a fixed prime and G is a Lannes group with finitely generated mod-p cohomology ring H^*BG. In this section we discuss results of [H2]. These results can be understood from the point of view developed in Section 4.1 but this would lead us too far afield, so we will only describe the results and try to explain their relevance for the study of group cohomology.

We will discuss in particular how much information about H^*BG is carried by the cohomology of various centralizers of elementary abelian p-subgroups of G. The centralizers will be indexed by suitable collections $\mathcal{O}$ of elementary abelian p-subgroups of G. These collections will have the following two properties:

- if $E \in \mathcal{O}$ and E' is conjugate to E then $E' \in \mathcal{O}$,
- if $E \in \mathcal{O}$ and $E \subset E'$ then $E' \in \mathcal{O}$.

We call such collections *open*. The justification for this terminology comes from the fact that the conjugacy classes of elementary abelian p-subgroups E are in $1 - 1$ correspondence with prime ideals $\mathfrak{p}_E \subset H^*BG$ which are invariant with respect to the reduced power operations P^i if $p > 2$ and with respect to all Sq^i if $p = 2$ [Q1]; the prime ideal $\mathfrak{p}_E$ is given as the radical of the kernel of the restriction map $H^*BG \longrightarrow H^*BE$. A homogeneous ideal which is invariant with respect to the Sq^i resp. P^i-operations will be called an *invariant ideal*.

Via this correspondence open collections correspond to intersections of the invariant part of the prime ideal spectrum of the graded ring H^*BG with subsets which are open in the Zariski topology. Furthermore, if $\mathfrak{a}$ is an ideal in H^*BG then it is clear that the collection

$$\mathcal{O}(\mathfrak{a}) := \{E \in \mathcal{A}_p(G)|\mathfrak{a} \not\subset \mathfrak{p}_E\}$$

forms an open collection, and this collection depends only on the radical of the ideal $\mathfrak{a}$. Conversely, if an open collection $\mathcal{O}$ is given, it determines a radical invariant ideal

$$\mathfrak{a}(\mathcal{O}) = \bigcap_{E \not\in \mathcal{O}} \mathfrak{p}_E$$

and the open subset in the Zariski topology is given as the set of prime ideals not containing $\mathfrak{a}$. The assignments $\mathfrak{a} \mapsto \mathcal{O}(\mathfrak{a})$ and $\mathcal{O} \mapsto \mathfrak{a}(\mathcal{O})$ establish bijections between the radical invariant ideals of H^*BG and open collections in $\mathcal{A}_p(G)$ (cf. [H2]).

For example, if $\mathfrak{a} = \widetilde{H}^*BG$ is the ideal consisting of the elements of positive degree in H^*G then $\mathcal{O}(\mathfrak{a})$ is precisely the collection of all non-trivial elementary abelian p-subgroups; if G is a compact Lie group, this is an immediate consequence of Theorem 3 in section 1, in the general case one can deduce it from Proposition 29 below.

For an open collection $\mathcal{O}$ we will denote the full subcategory of $\mathcal{A}_p(G)$ with object set $\mathcal{O}$ also by $\mathcal{O}$. The assignment $E \mapsto H^*BC_G(E)$ extends to a functor from $\mathcal{O}$ to unstable modules with an H^*BG-module structure (which for $H^*BC_G(E)$ is given via the restriction homomorphism).

Here is a preliminary version of the main result of this section. A stronger result in which we will also establish the link with suitable localizations will be given below.

Theorem 26 [H2]. *Let G be a Lannes group such that H^*BG is a finitely generated $\mathbb{F}_p$-algebra. Let $\mathfrak{a} \subset H^*BG$ be a radical invariant ideal and let $\mathcal{O}(\mathfrak{a})$ be the associated open collection of elementary abelian p-subgroups. Then the restriction maps induce a map*

$$\rho : H^*BG \longrightarrow \lim_{\mathcal{O}(\mathfrak{a})} H^*BC_G(E)$$

whose kernel and cokernel are $\mathfrak{a}$-torsion, i.e. they are annihilated by some large power of $\mathfrak{a}$.

Note that from a computational point of view the two properties defining an open collection are reasonable in the following sense: If $E \subset E'$ then the centralizer $C_G(E')$ will be contained in $C_G(E)$ and will usually be easier to understand than $C_G(E)$, both group theoretically and cohomologically. (In fact, $H^*BC_G(E')$ is determined by $H^*BC_G(E)$ via Lannes' Theorem applied to $H^*BC_G(E)$. Can you make this precise?) In order to apply the theorem in concrete cases one would try to compute as many $H^*BC_G(E)$ as possible; the minimal E's in that collection define an open collection $\mathcal{O}$ of elementary abelian p-subgroups (all those which contain one of the minimal ones) and chances are that $H^*BC_G(E)$ can be computed for all $E \in \mathcal{O}$ and with some luck $\lim_{\mathcal{O}(\mathfrak{a})} H^*BC_G(E)$ can be computed as well.

We discuss some special cases of Theorem 26.

1) Suppose E is a maximal elementary abelian p-subgroup. Then the collection of all elementary abelian p-subgroups which are conjugate to E form an open collection $\mathcal{O}(\mathfrak{a})$. We denote the corresponding radical invariant ideal $\mathfrak{a}$ by $\mathfrak{a}(E)$. This ideal consists of all elements which restrict nilpotently to E' whenever E' is not conjugate to E. If we abbreviate $\mathrm{Aut}_{A_p(G)}(E)$ by A then the inverse limit over the category $\mathcal{O}$ can be identified with the invariants $(H^*BC_G(E))^A \cong (H^*BC_G(E))^{N_G(E)}$ where $N_G(E)$ denotes the normalizer of E in G. Therefore Theorem 26 yields the following result.

Corollary 27 (cf. [Du2]). *Suppose G is a Lannes group such that H^*BG is a finitely generated $\mathbb{F}_p$-algebra. Let E be a maximal elementary abelian p-subgroup. Then the kernel and cokernel of the map*

$$\rho : H^*BG \longrightarrow (H^*BC_G(E))^{N_G(E)}$$

are torsion with respect to $\mathfrak{a}(E)$.   □

We have seen in Section 1 that there exists an element $z_E \in H^*BG$ which restricts trivially to all elementary abelian p-subgroups which are not conjugate to E and to a power of the top Dickson invariant on all elementary abelian p-subgroups which are conjugate to E. So z_E belongs to $\mathfrak{a}$ but not to $\mathfrak{p}_E$, and therefore $\mathfrak{a}$-torsion modules are killed by localization at $\mathfrak{p}_E$. Now localization at $\mathfrak{p}_E$ is exact and hence we derive the result of Duflot that $(H^*BG)_{\mathfrak{p}_E}$, the localization of H^*BG at the minimal invariant prime ideal $\mathfrak{p}_E$ (given as radical of the kernel of restriction to a maximal elementary abelian p-subgroup E) is isomorphic to $((H^*BC_G(E))_{\mathfrak{p}_E})^{N_G(E)}$, the invariants of the localization of $H^*BC_G(E)$ with respect to the action of $N_G(E)$. Corollary 27 can in this case be considered as a strengthening of Duflot's result in so far as it gives more precise information about the kernel and cokernel of the map ρ.

2) At the other extreme we consider the collection of all non-trivial elementary abelian p-subgroups of G. The corresponding category will be denoted $\mathcal{A}_p^*(G)$.

As remarked above the corresponding ideal $\mathfrak{a}$ is just the augmentation ideal $\widetilde{H}^*BG$. It is clear that in this case a finitely generated torsion module is nothing but a finite module. If G is a compact Lie group, then by Theorem 3 the cohomology of the centralizers $C_G(E)$ are finitely generated H^*BG-modules for all E. In fact, this remains true if G is any Lannes group for which H^*BG is a finitely generated $\mathbb{F}_p$-algebra [DW3], [H7]. It follows that the target of ρ is finitely generated as H^*BG-module. Therefore the kernel and cokernel of ρ are finitely generated $\mathfrak{a}$-torsion modules, hence finite; in particular ρ is an isomorphism in all sufficiently large degrees. In other words Theorem 26 specializes to the following result.

Corollary 28.   *Suppose G is a Lannes group such that H^*BG is a finitely generated $\mathbb{F}_p$-algebra. Then the kernel and cokernel of the map*

$$\rho : H^*BG \longrightarrow \lim_{\mathcal{A}_p^*(G)} H^*BC_G(E)$$

are finite.   □

   In the case of a compact Lie group more can be said: if G contains an element of order p then according to an important theorem of Jackowski and McClure [JM] ρ is an isomorphism (see Theorem 32 below). The reader is also encouraged to consult section 8 and 13 of Dwyer's notes (part I of this volume) where the case of finite groups is treated.

3) In the general case we can use Theorem 26 to control the *size* (by which we mean the Krull dimension, or equivalently the order of the pole of the Poincaré series at $t = 1$) of the kernel of the map ρ; in fact, the size of this kernel can be bounded by the size of $H^*BG/\mathfrak{a}$ and this can in turn be bounded by the maximal rank of an $E \notin \mathcal{O}$. Furthermore, the theorem also gives a tool to construct elements in H^*BG: for this one can take any element $z \in \lim_{\mathcal{O}(\mathfrak{a})} H^*BC_G(E)$; if $x_1, \cdots, x_k$ are in $\mathfrak{a}$ and k is large enough then $x_1 \cdots x_k z$ is in the image of the map. Non-trivial elements x_i can often be constructed as characteristic classes and bounds for k are also often available, e.g. via the methods of part II of [HLS3]. The existence of non-trivial classes $x_i \in \mathfrak{a}$ is also guaranteed by the following result.

Proposition 26 [CaH].   *Assume G is a Quillen group such that $\mathcal{A}_p(G)$ is equivalent to a finite category. Let $\mathcal{O} \subset \mathcal{A}_p(G)$ be an open collection. Then there is an element $x \in H^*BG$ which restricts nilpotently to H^*BE iff $E \notin \mathcal{O}$.*

Examples.
   a) In [H2] we used Theorem 26 to give an upper bound for the size of the kernel K_n of the restriction map from $H^*(BGL(n, \mathbb{Z}[1/2]); \mathbb{F}_2)$ to $H^*(BD_n; \mathbb{F}_2)$ (with D_n denoting the subgroup of diagonal matrices, i.e. $D_n \cong (\mathbb{Z}[1/2]^\times)^n \cong \mathbb{Z}^n \times (\mathbb{Z}/2)^n$). This kernel is non-trivial for n large, e.g. if $n = 32$ [Dw2]. If $\sigma(K_n)$ denotes this size, and if n_0 is the minimal

n with K_n non-trivial, then Theorem 26 can be used to show $\sigma(K_n) \leq n - n_0 + 1$ for $n \geq n_0$. In fact, we show even more in [H2], namely $\sigma(K_n) = n - n_0 + 1$ for $n \geq n_0$.

b) In joint work with F. Cohen and Y. Xia we will investigate certain mapping class groups and will produce new classes in their cohomology by the method outlined above.

Theorem 26 is contained in the following more general result. To state it we need to introduce more notation.

First we need to extend the context in which we are working. We consider the category $H^*BG - \mathcal{U}$ of unstable H^*BG-modules. Its objects are unstable modules M over the Steenrod algebra equipped with the structure of a graded H^*BG-module such that the structure map $H^*BG \otimes M \longrightarrow M$ is A_p-linear, i.e. it obeys the Cartan formula. Morphisms are degree preserving homomorphisms which are both A_p-linear and H^*BG-linear. The full subcategory whose objects are finitely generated as H^*BG-modules is denoted by $H^*BG_{fg} - \mathcal{U}$. Natural objects of $H^*BG - \mathcal{U}$ are given by H^*BG, $H^*BC_G(E)$ or more generally by the cohomology $H^*_G X := H^*(EG \times_G X)$ of the Borel construction of a G-space X. If X is a finite G-CW-complex and G is a suitable discrete group (e.g. if G is finite) then $H^*_G X$ is even in $H^*BG_{fg} - \mathcal{U}$.

For an invariant ideal $\mathfrak{a}$ and an unstable H^*BG-module M we consider

$$F_\mathfrak{a}M := \{x \in M | \mathfrak{a}^n x = 0 \text{ for some } n \in \mathbb{N}\} .$$

It is straightforward to check (using the Cartan formula) that $F_\mathfrak{a}$ defines a functor from $H^*BG - \mathcal{U}$ resp. $H^*BG_{fg} - \mathcal{U}$ to itself. This functor is left exact, and as the category $H^*BG_{fg} - \mathcal{U}$ is abelian with enough injectives [H2], $F_\mathfrak{a}$ has right derived functors denoted by $R^i F_\mathfrak{a}$. If $\mathfrak{a} = \widetilde{H^*BG}$ and M is in $H^*BG_{fg} - \mathcal{U}$ then it is clear that $F_\mathfrak{a}M$ is the largest finite subobject of M; in this case we denote the functor simply by F and its derived functors by $R^i F$. The existence of enough injectives in $H^*BG_{fg} - \mathcal{U}$ implies that for $M \in H^*BG_{fg} - \mathcal{U}$ all the modules $R^i FM$ are finite.

Here is the generalization of Theorem 26.

Theorem 30 [H2]. *Let G be a Lannes group with H^*BG a finitely generated algebra and $\mathfrak{a} \subset H^*BG$ an invariant ideal with associated open collection $\mathcal{O}(\mathfrak{a})$. Denote the full subcategory of $\mathcal{A}_p(G)$ with $\mathcal{O}(\mathfrak{a})$ as object set again by $\mathcal{O}(\mathfrak{a})$.*

a) *Then there is an exact sequence*

$$0 \longrightarrow F_\mathfrak{a}H^*BG \longrightarrow H^*BG$$
$$\xrightarrow{\rho} \lim_{\mathcal{O}(\mathfrak{a})} H^*BC_G(E) \longrightarrow R^1 F_\mathfrak{a}H^*BG \longrightarrow 0$$

in which the components of ρ are induced by the restriction maps. In particular, the kernel and cokernel of ρ are unstable finitely generated $\mathfrak{a}$-torsion modules. Furthermore, the map ρ is localization away from the subcategory of unstable finitely generated $\mathfrak{a}$-torsion modules.

b) *If $\lim^i$ denotes the i-th right derived functor of $\lim$ then there are natural isomorphisms*

$$\lim^i_{\mathcal{O}(\mathfrak{a})} H^* BC_G(E) \cong R^{i+1} F_{\mathfrak{a}} H^* BG$$

for all $i > 0$. In particular, $\lim^i_{\mathcal{O}(\mathfrak{a})} H^* BC_G(E)$ is an unstable finitely generated $\mathfrak{a}$-torsion module for all $i > 0$.

Comment on the proof. This result relies heavily on two results. Firstly, the category $H^* BG_{fg} - \mathcal{U}$ has enough injectives, and secondly these injectives are explicit enough (they are closely related to the injectives $H^* BV \otimes J(n)$ in the category $\mathcal{U}$) that for them one can check the theorem directly. The remainder of the proof is then standard homological algebra. □

The following special case of this theorem needs to be emphasized. In it $A_p^*(G)$ denotes the full subcategory of $A_p(G)$ consisting of all non-trivial elementary abelian p-subgroups of G.

Corollary 31 [H2]. *Let G be a Lannes group with $H^* BG$ a finitely generated algebra.*

a) *Then there is an exact sequence*

$$0 \longrightarrow FH^* BG \longrightarrow H^* BG$$
$$\overset{\rho}{\longrightarrow} \lim_{A_p^*(G)} H^* BC_G(E) \longrightarrow R^1 FH^* BG \longrightarrow 0$$

in which the components of ρ are induced by the restriction maps. In particular, the kernel and cokernel of ρ are finite. Furthermore, the map ρ is localization away from the subcategory of unstable finite $H^* BG$-modules.

b) *There are natural isomorphisms*

$$\lim^i_{A_p^*(G)} H^* BC_G(E) \cong R^{i+1} FH^* BG$$

for all $i > 0$. In particular, $\lim^i_{A_p^*(G)} H^* BC_G(E)$ is finite for all $i > 0$.

We emphasize that part (b) of this result is really relevant for the study of cohomology of groups. This will become clear in Section 6 below.

We close this section with some comments on and consequences of Theorem 30 and Corollary 31.

The results in [H2] are even more general; there is a version which holds for each unstable module $M \in H^* BG_{fg} - \mathcal{U}$ (in fact, $H^* BG$ can even be replaced by an unstable algebra K as long as it is finitely generated as an $\mathbb{F}_p$-algebra). In this generality the cohomology of the centralizers is replaced by certain "components" of $T_V M$; if $M = K = H^* BG$ these components are given via Theorem 19 by the cohomology of the spaces $BC_G(E)$. We refer the reader to [H2] for more details. Here we just want to remark that these "components" serve in the world of unstable $H^* BG$-modules as a replacement of the sheaf associated to a module over a commutative ring in the classical context. In particular, in the world of unstable modules and unstable algebras the functor $E \mapsto H^* BC_G(E)$ defined on $\mathcal{A}_p(G)$ plays the role of the structure sheaf of the ring $H^* BG$, and this yields a conceptual explanation why the cohomology of the centralizers plays such a crucial role in understanding $H^* BG$. This point of view is also supported by the formal analogy of Theorem 30 with results of Grothendieck on local cohomology in the classical situation of modules over a noetherian commutative ring R [Gr].

If G is a finite group (or a compact Lie group) which contains an element of order p then one can show [H2] that $R^i F H^* BG$ is trivial for all $i \geq 0$ and therefore the map ρ in Corollary 31 is an isomorphism and the higher limits $\lim^i_{\mathcal{A}_p^*(G)} H^* BC_G(E)$ vanish for all $i > 0$. Thus one recovers a result by Jackowski and McClure [JM] resp. by Dwyer and Wilkerson [DW1] (see Theorem 32 below) which plays a central role in the homotopy theory of classifying spaces and p-compact groups ([DW2], [JMO]).

Examples.

a) If G is discrete or profinite then ρ need no longer be an isomorphism. In [H4] the "profinite anologue" of Corollary 31 was used to compute the continuous mod-p cohomology in large dimensions of $GL(p-1, \mathbb{Z}_p)$ and the $(p-1)$-st "Morava stabilizer groups", i.e. the (p-Sylow subgroups of the) group of units in the maximal order of the division algebras over $\mathbb{Q}_p$ whose dimension is $(p-1)^2$ and whose Hasse invariant is $\frac{1}{p-1}$. For $p = 3$ additional arguments allowed to extend the calculation in large dimensions to one in all dimensions and in those cases ρ turned out not to be an isomorphism.

b) The case of $H^*(BSL(3, \mathbb{Z}[1/2]); \mathbb{F}_2)$ was considered in [H3] and [H5]: Corollary 31 was used in [H3] to carry out the calculation in large dimensions, and building on this the calculation was completed in [H5] (see Section 6 below); in this case the complete calculation showed that ρ happens to be an isomorphism again.

Remark. Theorem 30 implies that the restriction map

$$\mathrm{res}_{\mathcal{O}(\mathfrak{a})} : H^* BG \longrightarrow \prod_{E \in \mathcal{O}(\mathfrak{a})} H^* BC_G(E)$$

is injective if and only if $F_{\mathfrak{a}} H^* BG = 0$. In particular, if there exists a non-trivial element in $\mathfrak{a}$ which is not a zero-divisor then clearly $\mathrm{res}_{\mathcal{O}(\mathfrak{a})}$ is injective. This was first proved in [HLS3]. The following "converse" was proved in [CaH]: if $\mathrm{res}_{\mathcal{O}(\mathfrak{a})}$ is injective then there exists an element $x \in \mathfrak{a}$ which is not a zero-divisor; in fact, any element provided by Proposition 29 has this property.

We finish this section by remarking that Corollary 31 is somewhat reminiscent of results on Farrell cohomology of groups of finite virtual cohomological dimension. We will come back to this in the next section.

6. The centralizer spectral sequence

As before p is a fixed prime. In this section the coefficients need not necessarily be equal to $\mathbb{F}_p$ so we will no longer skip them from the notation.

The higher limits of Theorem 30 and Corollary 31 have also geometric significance; they occur as E_2-term in a homotopy colimit spectral sequence. The reader uncomfortable with homotopy colimits is encouraged to consult [JMO] for an introduction to the concept, or Dwyer's notes (part I of this volume), in particular sections 4,5 and 6. In the current section homotopy colimits will be primarily discussed to the extent that they produce interesting spectral sequences; they can be defined for functors F from a small category $\mathcal{C}$ to topological spaces and give a recipe to glue together the spaces $F(c)$, $c \in \mathcal{C}$, in a homotopy invariant way encoded by the category $\mathcal{C}$. Our first example is the functor $E \mapsto EG \times_G (G/C_G(E))$ from $\mathcal{A}_p(G)^{op}$ to spaces. Note that if E is the trivial subgroup this functor takes value BG.

First we recall the following result of Jackowski and McClure which "reconstructs" BG from the spaces $EG \times_G (G/C_G(E)) \simeq BC_G(E)$ with E nontrivial.

Theorem 32 [JM]. *Let G be a compact Lie group (which may be finite but is assumed to contain an element of order p). Then the natural map*

$$\pi : \mathrm{hocolim}_{\mathcal{A}_p^*(G)^{op}} EG \times_G G/C_G(E) \longrightarrow BG$$

induces an isomorphism in homology with coefficients in the p-local integers $\mathbb{Z}_{(p)}$.

Starting from the result of Jackowski and McClure the following generalization was obtained in [H3]. To state it we need to introduce some notation. Let G be a (topological) group and X a G-space. If H is a subgroup of G then X^H denotes the subspace of X fixed by all of H; the p-singular locus X_s is defined to be $X_s := \bigcup_g X^{<g>}$ where g runs through all elements of G of order p and $< g >$ denotes the subgroup generated by g. The G-space X gives rise to a functor $E \mapsto EG \times_{C_G(E)} X^E$ from $\mathcal{A}_p(G)^{op}$ to spaces.

Theorem 33 [H3]. *Assume G is a discrete group or a compact Lie group and X is any G-CW-complex. If G is discrete assume that the isotropy group of each $x \in X$ is finite. Then the natural map*

$$\pi : \mathrm{hocolim}_{\mathcal{A}_p^*(G)^{op}} \, EG \times_{C_G(E)} X^E \longrightarrow EG \times_G X_s$$

induces an isomorphism with coefficients in $\mathbb{Z}_{(p)}$.

Outline of the proof. We regard this result as a comparison theorem on functors from G-spaces to graded $\mathbb{F}_p$-vector spaces, say. So let F_1^G be the functor which associates to a G-space X the cohomology of $\mathrm{hocolim}_{\mathcal{A}_p^*(G)^{op}} \, EG \times_{C_G(E)} X^E$, and F_2^G the functor which associates to X the cohomology of $EG \times_G X_s$. We want to show that the natural transformation π^* between these functors is an isomorphism for certain G-spaces X. These functors are G-homotopy invariant, they turn arbitrary coproducts into products and associate to homotopy pushouts Mayer-Vietoris sequences. Therefore it suffices to show that π^* is an isomorphism for each orbit G/K which appears in these spaces. The crucial observation is now that both functors satisfy "Shapiro type" isomorphisms $F_i^G(G/K) \cong F_i^K(pt)$, and then we are done by Theorem 32. $\qquad\square$

Of primary interest for us is the resulting homotopy colimit spectral sequence [BK].

Corollary 34 [H3]. *Assume G is a discrete group or a compact Lie group and X is a G-CW-complex. If G is discrete assume that the isotropy group of each $x \in X$ is finite. Then for every $\mathbb{Z}_{(p)}$-module M there exists a first quadrant cohomological spectral sequence*

$$E_2^{s,t} = \lim_{\mathcal{A}_p^*(G)}^s H_{C_G(E)}^t(X^E; M) \Longrightarrow H_G^{s+t}(X_s; M) . \qquad\square$$

There are strong vanishing results for this spectral sequence in case $M = \mathbb{F}_p$. In fact, in [JM] Theorem 32 was proved by showing via Mackey functor techniques that in this case the spectral sequence is concentrated on the vertical edge and the natural map $H^*(BG; \mathbb{F}_p) \longrightarrow \lim_{\mathcal{A}_p^*(G)} H^*(BC_G(E); \mathbb{F}_p)$ is an isomorphism. Dwyer and Wilkerson [DW1] gave a proof of this using techniques from the theory of unstable modules and in [H2] still another such proof was given (cf. the discussion after Corollary 31 above). In the general case the results of [H2] (cf. Theorem 30 above) imply:

Theorem 35 [H3]. *Assume G is a discrete group or a compact Lie group and X is a G-CW-complex with finitely many equivariant cells. If G is discrete assume that the isotropy group of each $x \in X$ is finite. If coefficients are in $\mathbb{F}_p$ then $E_2^{s,*}$ is a finite graded $\mathbb{F}_p$-vector space for each $s > 0$, in particular $E_2^{s,t} = 0$ if $s > 0$ and t is large.*

An additional vanishing result has been proved by Oliver. (From the point of view of the discussion after Corollary 31 above this result corresponds to the vanishing of local cohomology above the Krull dimension of a ring.)

Theorem 36 [O]. *Let G be any group and let F be any functor from $\mathcal{A}_p^*(G)$ to $\mathbb{Z}_{(p)}$-modules. Then $\lim_{\mathcal{A}_p^*(G)}^s F = 0$ for all $s \geq r_p(G)$, the p-rank of G.*

As a consequence, under the assumptions of Theorem 35 and if $r_p(G)$ is finite, the spectral sequence is essentially concentrated on the vertical edge and it almost collapses. In particular, the edge homomorphism $H_G^*(X_s; \mathbb{F}_p) \longrightarrow \lim_{\mathcal{A}_p^*(G)} H_{C_G(E)}^*(X^E; \mathbb{F}_p)$ is an isomorphism in large dimensions.

These results have interesting consequences for computations of group cohomology. First we introduce certain classes of groups by slightly modifying a definition due to Kropholler [K].

Definition 8. A discrete group G is said to be an $\mathcal{H}_1\mathcal{F}$-group iff there exists a G-CW-complex X with the following properties:

- X is finite dimensional and mod-p acyclic.
- The isotropy group G_x of each $x \in X$ is a finite group.

  In this case we call X an $\mathcal{H}_1\mathcal{F}$-complex for G.

- If in addition $G\backslash X$ is compact, then G is said to be a $\mathcal{K}_1\mathcal{F}$-group and X a $\mathcal{K}_1\mathcal{F}$-complex for G.

In the appendix of [H2] we proved that a $\mathcal{K}_1\mathcal{F}$-group G is always a Lannes group and H^*BG is always a finitely generated $\mathbb{F}_p$-algebra. Extensions to certain $\mathcal{H}_1\mathcal{F}$-groups were discussed in the appendix of [H6].

As in [K] one can define $\mathcal{H}_\alpha\mathcal{F}$-groups for every ordinal α but we will refrain from doing so because we are only interested in the case $\alpha = 1$. All groups G of finite virtual cohomological dimension are $\mathcal{H}_1\mathcal{F}$-groups and very often they are even $\mathcal{K}_1\mathcal{F}$-groups, e.g. if G is an $(S$-$)$ arithmetic group [Se], a mapping class group [Ha], an outer automorphism group of a free group [CV] or a word hyperbolic group in the sense of Gromov [GH].

If X is a mod-p acyclic G-CW-complex then

$$H_G^*(X; \mathbb{F}_p) \cong H^*(BG; \mathbb{F}_p),$$

and if X is also finite dimensional then Smith theory implies that the fixed point sets X^E are also mod-p acyclic and therefore

$$H_{C_G(E)}^*(X^E; \mathbb{F}_p) \cong H^*(BC_G(E); \mathbb{F}_p).$$

Furthermore $H_G^*(X; \mathbb{F}_p) \cong H_G^*(X_s; \mathbb{F}_p)$ if $*$ exceeds the dimension of X; more precisely, the relative group $H_G^*(X, X_s; \mathbb{F}_p)$ is naturally isomorphic to $H^*(G\backslash(X, X_s); \mathbb{F}_p)$ and therefore vanishes if $*$ exceeds the dimension of X.

Thus we have an almost collapsing "centralizer spectral sequence" which converges not to $H^*(BG; \mathbb{F}_p)$ but to the closely related object $H^*_G(X_s; \mathbb{F}_p)$. More precisely we have the following result.

Corollary 37 [H3].

a) *Assume G is an $\mathcal{H}_1\mathcal{F}$-group, X an $\mathcal{H}_1\mathcal{F}$-complex for G and M is a $\mathbb{Z}_{(p)}$-module. Then there exists a first quadrant cohomological spectral sequence*

$$E_2^{s,t} = \lim^s_{\mathcal{A}^*_p(G)} H^t(BC_G(E); M) \Longrightarrow H_G^{s+t}(X_s; M) .$$

with $E_2^{s,*} = 0$ for $s \geq r_p(G)$.

b) *If G is a $\mathcal{K}_1\mathcal{F}$-group, X a $\mathcal{K}_1\mathcal{F}$-complex for G and $M = \mathbb{F}_p$ then $E_2^{s,*}$ is a finite graded $\mathbb{F}_p$-vector space for each $s > 0$ and $E_2^{s,*} = 0$ for $s \geq r_p(G)$. Furthermore, both the edge homomorphism*

$$H^*_G(X_s; \mathbb{F}_p) \longrightarrow \lim_{\mathcal{A}^*_p(G)} H^*(BC_G(E); \mathbb{F}_p)$$

and the map

$$\rho : H^*(BG; \mathbb{F}_p) \longrightarrow \lim_{\mathcal{A}^*_p(G)} H^*(BC_G(E); \mathbb{F}_p)$$

are isomorphisms in large dimensions.

The corollary suggests a program for computing the mod-p cohomology of a $\mathcal{K}_1\mathcal{F}$-group consisting of the following steps (each of which can be quite difficult though).

- Determine the (category of) elementary abelian p-subgroups of G and the centralizers of these subgroups.

- Compute the mod-p cohomology of (the classifying spaces of) the centralizers (as a functor on $\mathcal{A}^*_p(G)$).

- Compute $H^*_G(X_s; \mathbb{F}_p)$ from the (almost collapsing) centralizer spectral sequence. (Note that the input into this spectral sequence is independent of the choice of X!)

- Find a suitable $\mathcal{K}_1\mathcal{F}$-complex X, compute $H^*(G \backslash (X, X_s); \mathbb{F}_p)$ and evaluate the long exact mod-p cohomology sequence of the pair $EG \times_G (X, X_s)$. (This part is vital for getting at the low-dimensional cohomology of G.)

Example. The group $G = SL(3, \mathbb{Z}[1/2])$ is known to be a $\mathcal{K}_1\mathcal{F}$-group. For $M = \mathbb{F}_2$ this group provides a particularly favourable example for this program and was dealt with in [H3] and [H5].

The first three steps do not involve the space X explicitly and are relatively easy. They were carried out in [H3]; over $\mathbb{Z}[1/2]$ all elementary abelian 2-subgroups in $SL(n, \mathbb{Z}[1/2])$ are diagonalizable, the centralizers in case $n = 3$ are small and their cohomology was known. The 2-rank is 2, so the spectral sequence has only two columns ($s = 0, 1$) and degenerates into a short exact sequence

$$0 \longrightarrow \Sigma \lim^1_{\mathcal{A}^*_p(G)} H^*(BC_G(E); \mathbb{F}_2) \longrightarrow H^*_G(X_s; \mathbb{F}_2) \longrightarrow$$

$$\lim_{\mathcal{A}^*_p(G)} H^*(BC_G(E); \mathbb{F}_2) \longrightarrow 0 .$$

Here Σ denotes the suspension functor and appears here in order to make all maps degree preserving. The inverse limit turns out to be isomorphic to $\mathbb{F}_2[w_2, w_3] \otimes \Lambda(e_3, e_5)$, the tensor product of a polynomial algebra on the universal Stiefel Whitney classes w_2 and w_3 (coming from the embedding of $\mathbb{Z}[1/2] \subset \mathbb{R}$) and an exterior algebra on classes e_3 and e_5 of dimension 3 and 5 respectively. The $\lim^1$-term is non-trivial; it is isomorphic to $\Sigma^3 \mathbb{F}_2$.

The last step of the program was carried out in [H5]; this was quite involved and required a careful analysis of the action of $SL(3, \mathbb{Z}[1/2])$ on a suitable space X; such a space is given by the product $X_2 \times X_\infty$ where X_2 denotes the Bruhat-Tits-building for $SL(3, \mathbb{Q}_2)$ and X_∞ the symmetric space $SL(3, \mathbb{R})/SL(3, \mathbb{Z})$. The relative cohomology turned out to compensate the contribution coming from the $\lim^1$-term and the final result can be stated as follows.

Theorem 38 [H5]. *The map ρ induces an isomorphism*

$$H^*(BSL(3, \mathbb{Z}[1/2]); \mathbb{F}_2) \cong \mathbb{F}_2[w_2, w_3] \otimes \Lambda(e_3, e_5) .$$

Furthermore the restriction map to the diagonal matrices in $SL(3, \mathbb{Z}[1/2])$ is injective.

Example. Another interesting application of Theorem 33 (which does not strictly follow the program outlined above) occurs in joint work in progress with J. Lannes. We study the mod-2 cohomology of the orthogonal group (with respect to the standard bilinear form) $O(n, \mathbb{Z}[1/2])$ and show that the reduction map to $O(n, \mathbb{F}_3)$ induces an isomorphism in mod-2 cohomology for $n \leq 14$. The groups $O(n, \mathbb{Z}[1/2])$ belong to the class $\mathcal{K}_1 \mathcal{F}$ and in the range $n \leq 14$ we show that $X = X_s$ for a suitable "building" which serves as a $\mathcal{K}_1 \mathcal{F}$-complex for $O(n, \mathbb{Z}[1/2])$. This together with Theorem 33 enables us to prove by induction that the reduction map induces an isomorphism for $n \leq 14$.

Similarly, if $O'(n, \mathbb{Z}[1/2])$ and $O'(n, \mathbb{F}_3)$ denote the orthogonal groups for the bilinear form $(x, y) \mapsto \Sigma_{i<n} x_i y_i + 2 x_n y_n$, then the reduction map induces an isomorphism for $n \leq 13$ but not for $n = 14$. This can be used to show that the restriction map in mod-2 cohomology from $GL(n, \mathbb{Z}[1/2])$ to its subgroup D_n of diagonal matrices is not injective for $n \geq 14$ thus improving on Dwyer's result [Dw2] which required $n \geq 32$ (see Example (a) after Proposition 29).

The "centralizer spectral sequence" has certain similarities with a "normalizer spectral sequence" in Farrell cohomology for groups of finite virtual cohomological dimension. Recall that above the virtual cohomological dimension Farrell cohomology (with any coefficients) agrees for such groups with ordinary cohomology. Furthermore there is a spectral sequence whose E_1-term is given by the Farrell cohomology of intersections of normalizers of elementary abelian p-subgroups of G and which converges to the Farrell cohomology of G [Br]. The relation between the two constructions leading to these spectral sequences has been investigated in work of [Sl] and more recently [Dw1] (if G is finite). The input in both spectral sequences is different; in the case of the centralizer spectral sequence it is given in explicit homological terms (namely as derived functors of lim) and we believe that it is therefore easier to handle. The example of $SL(3, \mathbb{Z}[1/2])$ and $p = 2$ gives some evidence to this claim.

Given that the two spectral sequences abut to the same object (at least in large dimensions) one starts wondering whether they might agree from some stage on. There is a bit of experimental evidence in this direction. E.g. in the case of $SL(3, \mathbb{Z}[1/2])$ the two spectral sequences agree in large total degree from E_2 on. Results by Chun Nip Lee [Le] on the edge homomorphism of the spectral sequence give some more evidence.

We summarize by expressing our belief that the "centralizer spectral sequence" is more effective for computing cohomology with trivial coefficients $\mathbb{F}_p$. Its E_2-term is more conceptual and its abutment is more geometrical (the cohomology of the Borel construction $EG \times_G X_s$ vs. the Farrell cohomology whose geometrical interpretation in low dimension is less clear). In fact, this geometric aspect was quite useful in computing the low dimensional cohomology of $SL(3, \mathbb{Z}[1/2])$ [H5].

In the case of a coefficient module M with non-trivial G-action the centralizer spectral sequence still converges to $H^*(BG; M)$ in sufficiently large dimensions [Dw3]. However, it is not clear whether there are still vanishing results for the E_2-term in the spirit of Theorem 35.

References

[AGM] J. F. Adams, J. H. C. Gunawardena and H. R. Miller, *The Segal conjecture for elementary abelian p-groups*, Topology **24** (1985), 435–460.

[AM]   A. Adem and R. J. Milgram, *Cohomology of finite groups*, Grundlehren der mathematischen Wissenschaften 309, Springer Verlag, 1994.

[AC]   G. S. Avrnunin and J. F. Carlson, *Nilpotency degree of cohomology rings in characteristic* 2, Proc. AMS **118** (1993), 339–343.

[Be]   D. Benson, *Representations and cohomology II: Cohomology of groups and modules*, Cambridge studies in advanced mathematics 31, Cambridge University Press, 1991.

[BK]   A. K. Bousfield and D. M. Kan, *Homotopy limits, completions and localizations*, Lecture Notes in Math. 304 Springer, 1972.

[Br]   K. Brown, *Groups of virtually finite dimension*, London Math. Soc. LNS 36, Homological Group Theory (edited by C.T.C. Wall), Cambridge University Press, 1979, pp. 27–70.

[CaH] J. F. Carlson and H.-W. Henn, *Cohomological detection and regular elements in group cohomology*, Proc. Amer. Math. Soc. **124** (1996), 665–670.

[C]   G. Carlsson, *G. B. Segal's Burnside ring conjecture for* $(\mathbb{Z}/2)^k$, Topology **22** (1985), 83–103.

[CV]   M. Culler and K. Vogtmann, *Moduli of graphs and automorphisms of free groups*, Invent. math. **84** (1986), 91–119.

[Du1] J. Duflot, *Smooth toral actions*, Topology **284** (1983), 253–265.

[Du2] J. Duflot, *Localization of equivariant cohomology rings*, Trans. Amer. Math. Soc. **284** (1984), 91–105.

[Dw1] W.G. Dwyer, *Sharp homology decompositions for classifying spaces of finite groups*, Group Representations: Cohomology, group actions and topology (Seattle, WA 1996). Proceedings of Symposia in Pure Mathematics **63** (1998), 197–220.

[Dw2] W.G. Dwyer, *Exotic cohomology for* $GL_n(\mathbb{Z}[1/2])$, Proc. Amer. Math. Soc. **126** (1998), 2159–2167.

[Dw3] W.G. Dwyer, *private comunication*.

[DW1] W. Dwyer and C. Wilkerson, *A cohomology decomposition theorem*, Topology **31** (1992), 433–443.

[DW2] W. G. Dwyer and C. W. Wilkerson, *Homotopy fixed point methods for Lie groups and finite loop spaces*, Annals of Mathematics **139** (1994), 395–442.

[DW3] W. Dwyer and C. Wilkerson, *Smith theory and the functor T*, Comm. Math. Helv. **66** (1991), 1–17.

[E1]   L. Evens, *The cohomology ring of a finite group*, Trans. AMS **101** (1961), 224–239.

[E2]   L. Evens, *The cohomology of groups*, Oxford Mathematical Monographs, Clarendon Press, 1991.

[Ga]   P. Gabriel, *Des catégories abéliennes*, Bull. Soc. Math. Fr. **90** (1962), 323–448.

[Gr]   A. Grothendieck (notes by R. Hartshorne), *Local cohomology*, Springer Lecture Notes in Math. 41, 1967.

[GH]   E. Ghys and P. de la Harpe, *Sur les Groupes Hyperboliques d'après Mikhail Gromov*, Progress in Mathematics 83, Birkhäuser, 1990.

[GLZ]   J. Gunawardena, J. Lannes and S. Zarati, *Cohomologie des groupes symétriques et application de Quillen*, Advances in Homotopy Theory, London Math. Soc., Lect. Notes Series **139**, 61–68.

[Ha]   J. Harer, *The virtual cohomological dimension of the mapping class group of an orientable surface*, Invent. math. **84** (1986), 157–176.

[H1]   H.-W. Henn, *Classifying spaces with injective mod-p cohomology*, Comm. Math. Helv. **64** (1989), 200–206.

[H2]   H.-W. Henn, *Commutative algebra of unstable K-modules, Lannes' T-functor and equivariant mod-p-cohomology*, Journal für die reine und angewandte Mathematik **478** (1996), 189–215.

[H3]   H.-W. Henn, *Centralizers of elementary abelian p-subgroups, the Borel construction of the singular locus and applications to the cohomology of discrete groups*, Topology **36** (1997), 271–286.

[H4]   H.-W. Henn, *Centralizers of elementary abelian p-subgroups and mod-p cohomology of profinite groups*, Duke Mathematical Journal **91** (1998), 561–585.

[H5]   H.-W. Henn, *The cohomology of SL(3, ℤ[1/2])*, K-theory **16** (1999), 299–359.

[H6]   H.-W. Henn, *Unstable modules over the Steenrod algebra and cohomology of groups*, Group Representations: Cohomology, group actions and topology (Seattle, WA 1996). Proceedings of Symposia in Pure Mathematics **63** (1998), 277–300.

[H7]   H.-W. Henn, *Finiteness Properties of injective resolutions of certain unstable modules over the Steenrod algebra and applications*, Math. Ann. **291** (1991), 191–203.

[HLS1]   H.-W. Henn, J. Lannes and L. Schwartz, *Analytic functors, unstable algebras and cohomology of classifying spaces*, Contemp. Mathematics **96** (1989), 197–220.

[HLS2]   H.-W. Henn, J. Lannes and L. Schwartz, *The categories of unstable modules and unstable algebras modulo nilpotent objects*, American Journal of Math. **115** (1993), 1053–1106.

[HLS3]   H.-W. Henn, J. Lannes and L. Schwartz, *Localization of unstable A-modules and equivariant mod-p cohomology*, Math. Ann. **301** (1995), 23–68.

[IK]   K. Inoue and A. Kono, *Nilpotency of a kernel of the Quillen map*, J. Math. Kyoto Univ. **33** (1993), 1047–1055.

[JM]   S. Jackowski and J. McClure, *Homotopy decomposition of classifying spaces via elementary abelian subgroups*, Topology **31** (1992), 113–132.

[JMO]   S. Jackowski, J. McClure and R. Oliver, *Homotopy theory of classifying spaces of compact Lie groups*, Algebraic topology and its applications, Springer-Verlag, 1994, pp. 81–123.

[K]   P. Kropholler, *On groups of type* $(FP)_\infty$, Journal of pure and applied algebra **90** (1993), 55–67.

[Ln1]   J. Lannes, *Sur la cohomologie modulo p des p-groupes abéliens élémentaires*, Proceedings of the Durham Symposium on Homotopy theory 1985, Cambridge University Press 1987.

[Ln2]   J. Lannes, *Cohomology of groups and function spaces*, Preprint 1986.

[Ln3]   J. Lannes, *Sur les espaces fonctionnels dont la source est le classifiant d'un p-groupe abélien élémentaire*, Publ. Math. IHES **75** (1992), 135–244.

[LS]   J. Lannes and L. Schwartz, *Sur la structure des A-modules instables injectifs*, Topology **28** (1989), 153–169.

[LZ]   J. Lannes and S. Zarati, *Sur les U-injectifs*, Ann. Scient. Ec. Norm. Sup. **19** (1986), 303–333.

[Lz]   M. Lazard, *Groupes analytiques p-adiques*, Publ. Math. IHES **26** (1965).

[Le]   C. N. Lee, *Farrell cohomology and centralizers of elementary abelian p-subgroups*, Math. Proc. Cambridge Philos. Soc. **119** (1996), 403–417.

[M]   H. Miller, *The Sullivan Conjecture on maps from classifying spaces*, Ann. of Math. **120** (1984), 39–87.

[O]   R. Oliver, *Higher limits via Steinberg representations*, Comm. in Algebra **22** (1994), 1381–1393.

[Q1]   D. Quillen, *The spectrum of an equivariant cohomology ring I, II*, Ann. of Math. **94** (1971), 549–572, 573–602.

[Q2]   D. Quillen, *On the cohomology and K-theory of the general linear groups over a finite field* Ann. of Math. **96** (1972), 552–586.

[Sc]   L. Schwartz, *Unstable modules over the Steenrod algebra and Sullivan's fixed point set conjecture*, University of Chicago Press, 1994.

[Se]   J. P. Serre, *Arithmetic groups*, London Math. Soc. LNS 36, Homological Group Theory (edited by C.T.C. Wall), Cambridge University Press, 1979, pp. 105–136.

[Sl]   J. Slominśka, *Homotopy colimits on E-I-categories*, Algebraic Topology, Poznań 1989 Proceedings, Springer Lecture Notes in Math. 1474, 1991.

[SE]   N. Steenrod and D. B. A. Epstein, *Cohomology operations*, Ann. of Math. Studies 50, Princeton University Press, 1962.

[V]   B. B. Venkov, *Cohomology algebras for some classifying space*, Dokl. Akad. Nauk SSSR **127** (1959), 943–944.

Département de Mathématique
Institut de Recherche Mathématique Avancée
C.N.R.S. – Université Louis Pasteur
7 rue René Descartes
F-67084 Strasbourg
France